About This Report

C000061965

The U.S. Space Force faces adversaries that have demonstrated increasingly effective counterspace capabilities. Assuring mission success in a space warfighting domain means that the relevant capability has to be available ahead of threats, and it has to work in and through contested space environments. To respond to this challenge, the U.S. Space Force realizes that it must deliver capabilities of its own at an increasingly rapid pace. Traditional space system acquisition tends to operate on long, relatively inflexible life cycles that cannot keep up with short time frames or changes in needed capabilities. Thus, programs are increasingly pursuing alternative—rapid—pathways; however, some evidence suggests that this pursuit of speed may carry heretofore unidentified risks to mission assurance. The objective of this project was to identify critical risks to mission assurance created by rapid acquisition, to assess the potential impacts of these risks, and to recommend possible mitigations.

The research reported here was commissioned by the U.S. Space Force, Chief of Space Operations, and conducted by the Resource Management Program of RAND Project AIR FORCE from October 2020 to September 2021.

RAND Project AIR FORCE

RAND Project AIR FORCE (PAF), a division of the RAND Corporation, is the Department of the Air Force's (DAF's) federally funded research and development center for studies and analyses, supporting both the United States Air Force and the United States Space Force. PAF provides the DAF with independent analyses of policy alternatives affecting the development, employment, combat readiness, and support of current and future air, space, and cyber forces. Research is conducted in four programs: Strategy and Doctrine; Force Modernization and Employment; Resource Management; and Workforce, Development, and Health. The research reported here was prepared under contract FA7014-16-D-1000.

Additional information about PAF is available on our website:
www.rand.org/paf/

This report documents work originally shared with the DAF on September 27, 2021. The draft report, issued on September 30, 2021, was reviewed by formal peer reviewers and DAF subject-matter experts.

Acknowledgments

We would like to thank the many members of the U.S. Space Force and the U.S. Air Force, along with others at the U.S. Department of Defense and the Aerospace Corporation, who were willing to take the time to talk to us and provide useful inputs into the report. All participants spoke on a non-attribution basis, and although we cannot mention them by name, we are grateful for their inputs. Special thanks go to Christopher Ayres, Brig Gen Timothy Sejba, Robert Eidsmoe, Col Ben Brining, and Lt Col Gregg Izdepski for providing feedback throughout the project and helping us connect with key subject-matter experts. We would also like to thank Stephanie Young, Patrick Mills, and Anu Narayanan for Resource Management Program leadership and management, and for useful feedback throughout. Bill Shelton and Bonnie Triezenberg provided useful inputs during the research process, which helped inform the team. Finally, we are grateful to our peer reviewers, Philip Anton (Stevens Institute of Technology), Lauren Mayer (RAND) and Geoff Reber (Aerospace), whose contributions notably strengthened the report.

Summary

Issue

The U.S. Space Force (USSF) faces potential adversaries that have demonstrated increasingly effective counterspace capabilities. To outpace these threats, the USSF is pursuing rapid acquisition of warfighting capabilities. A key question is whether the acceleration of acquisition by the USSF using various techniques introduces any critical new risks. In particular, do the adaptations and streamlining techniques being used to get new space systems to operators quickly create vulnerabilities and challenges to mission assurance (MA) (i.e., the ability of operators to achieve their mission, continue critical processes, and protect people and assets in any operating environment or conditions)?[1]

The project was guided by the following questions:

- What *streamlining* techniques are being used to accelerate USSF acquisition?
- What potential *risks* are associated with those streamlining techniques?
- What is the potential impact of these streamlining techniques on *mission assurance*?
- What are potential *mitigations*?

Approach

We used a mixed methods approach to address the questions, including a review of government policies and literature on acquisition; discussions with over 40 subject-matter experts from the USSF, the Department of the Air Force, and federally funded research and development centers (FFRDCs); identification of potential sources of risk; creation of a framework for identifying the relative risk to MA of various events; identification of potential mitigation strategies; and analysis of Department of the Air Force data to identify common issues in programs using rapid acquisitions strategies.

Key Findings

- Streamlining methods across Space Systems Command (SSC) and the Space Rapid Capabilities Office (RCO) share some similarities, but differences are also evident,

---

[1] We adapted the definition provided in 2012 DoD Mission Assurance Strategy, which is "a process to protect or ensure the continued function and resilience of capabilities and assets—including personnel, equipment, facilities, networks, information and information systems, infrastructure, and supply chains—critical to the performance of DoD MEFs [mission-essential functions] in any operating environment or condition." Note that there are many definitions of *mission assurance*, and there is some overlap between our definition and the concept of space mission assurance (which includes resilience) described in Joint Publication 3-14, *Space Operations*.

driven by the urgency of the threat, complexity, organizational and structural resources, and risk tolerance of missions and culture.

- There are a series of critical risks that need to be addressed by USSF leadership across all rapid acquisition efforts:
  - insufficient alignment and coordination between the acquisition and operations communities
  - unreliable or inadequately timed financial resources
  - a shortage of on-site cybersecurity experts and intelligence personnel colocated with program offices
  - a lag in development of needed test capabilities and infrastructure
  - challenges in aligning software development life cycles
  - failure to consider and plan for systems evolution
  - alternative requirements processes that might specify capabilities that cannot be acquired on a rapid schedule.

- The programs using streamlining at SSC are still in the early stages of their life cycles and have not delivered products. Thus, MA outcomes of streamlining are not yet measurable.
- MA has traditionally focused on managing technical risk of the individual program, but MA for rapid acquisition should consider trade-offs between mission capability, reliability, resilience, security, and schedule to ensure mission success.

Table S.1. Key Differences in Mission Assurance Approach Between Traditional and Rapid Acquisition Programs

MA for Traditional Space Acquisition	MA for Rapid Space Acquisition
• Focuses on system	• Focuses on warfighter/mission
• Addresses technical risks to the narrow system	• Addresses technical, operational, and programmatic risks of the broader mission
• Averse to technical risk	• Tolerant of technical risk
• Maximizes performance-centric MA objectives (mission capability and reliability) that drive cost and schedule	• Balances multiple MA objectives (schedule, mission capability, reliability, security, resilience) within cost constraints

Recommendations

To address the above findings, we have identified some key actions for the USSF leadership:

- Expand the MA objectives for rapid acquisition to reflect the addition of new operational and programmatic goals on top of technical system goals.
- Address the risks associated with rapid acquisition identified above (see mitigation options in Chapter 3).
- Ensure that processes across the USSF acquisition and operational communities are updated to address the need to onboard capabilities more quickly. As these issues cross organizational boundaries, the acquisition community cannot address all of the challenges

itself, so other communities including the requirements and financial management will also need to make some changes.

- Proactively manage risks to MA associated with rapid acquisition by using the risk assessment framework and management process in Chapter 5 to provide a structured way to conceptualize MA from program inception; provide an approach for making intelligent risk trade-offs and choosing courses of action that ensures mission success; and offer an approach to manage risks collectively rather than individually.

Contents

Figures, Text Box, and Tables

Figures

Text Box

Tables

1. Introduction

The United States faces potential adversaries that have demonstrated increasingly effective counterspace capabilities. This is not a particularly recent challenge, although it has been growing over time. In January 2007, China shot down one of its own satellites,[2] sending a clear message about its capabilities. More recently, in July 2020, there were numerous news reports about Russia testing a space-based anti-satellite weapon, and a similar incident was reported in 2017.[3] The increasing threat in space was a key factor in the instantiation of the U.S. Space Force (USSF) as a separate and distinct military service uniquely focused on space as a warfighting domain. And the threat may necessitate agile, rapid responses that ensure that space guardians have the necessary capabilities in a timely way to effectively meet the threat.[4]

In the first year of the USSF's establishment, the Chief of Space Operations (CSO) outlined several priorities to guide the design of the new service, one of which was to "deliver new capabilities at operationally relevant speeds."[5] Not being able to operate effectively in the face of these new threats is an operational risk, and rapid acquisition and the delivery of new capabilities is part of the solution. That said, rapid processes may introduce new risks. Any effective response to the threat requires that the systems are not only available when needed, but that they also function as necessary and are robust against any challenge. The term *mission assurance* (MA) encompasses these last concepts. MA can be more formally defined as operators achieving the mission, continuing critical processes, and protecting people and assets in any operating environment or conditions.[6] But how do the need for acquisition speed and tools used to

---

[2] William J. Broad and David E. Sanger, "Flexing Muscle, China Destroys Satellite in Test," *New York Times*, January 18, 2007.

[3] Robert Burns, "US Accuses Russia of Testing Anti-Satellite Weapon in Space," *Washington Post*, July 23, 2020.

[4] Separate RAND work offers recommendations for USSF acquisition: William Shelton, Cynthia R. Cook, Charlie Barton, Frank Camm, Kelly Elizabeth Eusebi, Diana Gehlhaus, Moon Kim, Yool Kim, Megan McKernan, Sydne Newberry, and Colby P. Steiner, *A Clean Sheet Approach to Space Acquisition in Light of the New Space Force*, Santa Monica, Calif.: RAND Corporation, RR-A541-1, 2021.

[5] 1st Chief of Space Operations, *Chief of Space Operations' Planning Guidance*, November 9, 2020; Shirley Kan, *China's Anti-Satellite Weapon Test*, Congressional Research Service, RS22652, April 23, 2007.

[6] This definition is an adaptation of the definition provided in 2012 U.S. Department of Defense *Mission Assurance Strategy*, which is

> a process to protect or ensure the continued function and resilience of capabilities and assets— including personnel, equipment, facilities, networks, information and information systems, infrastructure, and supply chains—critical to the performance of DoD MEFs [mission-essential functions] in any operating environment or condition.

Note that there are many definitions of mission assurance, and there is some overlap between our definition and the concept of space mission assurance (which includes resilience) described in Joint Publication 3-14.

accelerate acquisition affect MA, as it has traditionally been conceived and implemented?[7] And if that impact might involve significant mission risk, what needs to be done to identify and mitigate this risk, starting at early stages of acquisition planning?

The traditional acquisition process for major capabilities, as described in the Department of Defense 5000 series of policies,[8] has a significant focus on risk reduction processes to ensure that capabilities as delivered meet the requirement. Unfortunately, this can mean that as much as a decade can pass between the identification of requirements and the delivery of new capabilities.[9] The increasing capabilities of potential adversaries and evolving offerings of the defense industrial base mean that the capabilities as delivered might no longer be adequate—or timely enough—to meet the threat. Delays in acquisition may themselves create significant MA challenges. At the same time, new approaches to acquisition—whether via streamlined processes or steps skipped or conducted concomitantly—may create unanticipated problems. Thus, the challenge is a balancing act: How can the USSF take advantage of rapid acquisition so that warfighters have the capabilities they need, without having to deal with new and unforeseen risks to MA? This is the fundamental question that we hoped to address in this project.

Streamlining Acquisition to Deliver Capabilities Faster

Focusing on the necessity of getting operationally relevant capabilities to warfighters in a timely way, the U.S. Department of Defense (DoD) has developed alternatives to traditional major capability acquisition, including those outlined in the Adaptive Acquisition Framework (AAF). Prior to the AAF, DoD relied on tailoring a set of acquisition processes. DoD also acquired a lot of smaller items through urgent needs processes. The AAF is attempting to reinforce tailoring as the default for programs. Several small, agile organizations have been set up within the Department of the Air Force (DAF) to focus on rapid acquisition, including the Rapid Capabilities Office (RCO) and the newer Space Rapid Capabilities Office (Space RCO). The need to ensure that acquisition of space capabilities occurs at the pace necessary to meet the threat is also the role of the Space Systems Command (SSC), the main USSF organization acquiring space capabilities. SSC uses accelerated acquisition approaches on a number of its programs.[10]

---

[7] As we discuss in Chapter 4, traditional MA is focused on technical and engineering aspects of the acquired system to assure with high confidence that the system will meet high performance and reliability requirements.

[8] Department of Defense Directive 5000.01, *The Defense Acquisition System*, September 9, 2020; Department of Defense Instruction 5000.02, *Operation of the Adaptive Acquisition Framework*, January 23, 2020.

[9] See, for example, Aerospace Corporation, *Pre-Contract Award Study Schedule Study*, TOR-2016-01191, 2016; Aerospace Corporation, *Why Does It Take So Long?* TOR-2018-00183, 2018.

[10] While Space RCO and SSC were the main focus of this analysis, we acknowledge that additional analysis is needed of space programs within Space Development Agency, Missile Defense Agency, and the National Reconnaissance Office. Each of these agencies is facing similar pressures and have comparable acquisition responsibilities as SSC and Space RCO. Inclusion of these other agencies would be beneficial in future analysis.

Accelerating the acquisition process can involve tailoring at any of the steps.[11] Efforts may include limited or other alternative approaches to testing, reduced management reviews, changes to intelligence inputs, limits on technical data required for approval of rapid procurement projects, and conducting the steps in parallel instead of sequentially. However, each of these decisions may create risks for the delivery, operation, or sustainment of needed capability to the warfighter, and thus affect MA. This is a particular concern when efforts supported by the rapid procurement are in mission critical, high-risk areas that may be targeted by adversaries, including, for example, satellites in low earth and geostationary orbit, military satellite communications (SATCOM) networks, cybersecurity for military payloads and other system elements, ground stations for missile warning systems and missile warning data processing, and other capabilities.

Thus, the driving question for this analysis is whether the acceleration of acquisition by these organizations using streamlining approaches introduces any critical new risks (even as they address the risk of not having operationally relevant systems when needed). In particular, do the adaptations and streamlining techniques being used to get new programs to operators quickly create vulnerabilities and challenges to MA? And if rapid acquisition is introducing risks to MA, how are these identified and addressed?

The main research challenge in addressing this question is that the current approaches to rapid acquisition in the USSF involve relatively new organizations (Space RCO) or processes (AAF pathways). Therefore, our analysis can only identify *potential* risks to MA, rather than those that have been encountered in operations. The Space RCO is a new organization, so any delivered systems would not have a long operational history. The non–major capability programs at SSC that are using other approaches in the AAF have not yet fielded systems. While there are examples of rapid space acquisition in other contexts, our focus on the new USSF means we focused on understanding potential impacts from these new approaches.

Research Questions and Approach

Our task was to assess whether pursuing streamlined approaches to acquisition is linked to critical risks to USSF MA (including system vulnerability), to assess the potential impacts of these risks, and to recommend possible mitigations. In the absence of a set of delivered or completed rapid programs with measurable MA to compare to traditional programs, we used a bottom-up approach to address this challenge and to answer these questions:

- What *streamlining* techniques are being used to accelerate USSF acquisition?

---

[11] Philip S. Anton, Brynn Tannehill, Jake McKeon, Benjamin Goirigolzarri, Maynard A. Holliday, Mark A. Lorell, and Obaid Younossi, *Strategies for Acquisition Agility: Approaches for Speeding Delivery of Defense Capabilities*, Santa Monica, Calif.: RAND Corporation, RR-4193-AF, 2020; Megan McKernan, Jeffrey A. Drezner, and Jerry M. Sollinger, *Tailoring the Acquisition Process in the U.S. Department of Defense*, Santa Monica, Calif.: RAND Corporation, RR-966-OSD, 2015.

- What potential *risks* are associated with those streamlining techniques?
- What is the potential impact of these streamlining techniques on *mission assurance*?
- What are potential *mitigations*?

We used a mixed methods approach to address these questions. We started with a review of relevant government policies and other literature on acquisition, including government documents. We then developed a set of questions and tailored them to use in semistructured interviews to elicit information in over 40 discussions with subject-matter experts (SMEs) as listed in Table 1.1. Using the literature review and interviews, we developed a compendium of potential risks to MA and created a framework for identifying potential mitigation strategies. We also conducted an analysis of DAF Monthly Acquisition Report (MAR) program reporting data to identify common issues identified in programs that are using rapid acquisition strategies.

Table 1.1. Space Organizations Interviewed

 Space RCO

★ Leadership
★ Contracting
★ Director's Action Group
★ Business Intelligence
★ Strategic Capabilities Group

 Space Systems Center

+ Development Corps, Strategic Systems Division
+ Enterprise Corps
+ Atlas Corps
+ Production Corps
+ Capability Integration, Portfolio Architecture
+ Prototype Operations
+ Innovation and Prototyping Directorate
+ Contracting
+ Cross Mission Ground & Communications Directorate
+ Innovation and Prototyping Directorate and Commander
+ Engineering & Mission Design Division

Other interviews: DAF RCO, Operator community, Aerospace, SDA, SpOC

Clarification of Terms

The terms *urgent* and *rapid* have been applied to acquisition in several ways since 1994 (see Appendix A for a history of rapid acquisition methods). In addition, the terms *risk* and *mission assurance*, which are also key parts of this analysis, have varying definitions. To ensure consistency of use throughout the analysis, we identified formal definitions of *urgent, rapid, risk,* and *mission assurance*, which we provide in Table 1.2.

In the context of space acquisition, MA is traditionally referred to as a function or a process and defined as "the disciplined application of general systems engineering, quality, and management principles towards the goal of achieving mission success," per the Aerospace

Corporation's guidebook on MA.[12] As this classic definition of MA indicates, MA has been focused on the system being acquired, as it typically means that the system will function as intended throughout the system's mission lifetime. However, we assert that in a contested space environment, the definition of MA needs to be broadened to focus on the warfighting mission, as shown in Table 1.2.

Table 1.2. Clarification of Critical Terms Used in this Analysis

Term	Definition
Urgent	• Refers to urgent operational needs and other quick reaction capabilities that can be fielded in less than two years using the Urgent Capability Acquisition Pathway
Rapid	• Refers to capabilities that have a level of maturity to allow them to be rapidly prototyped within an acquisition program or rapidly fielded, within five years • Also refers to rapid and iterative delivery of software capability (e.g., software-intensive systems and/or software-intensive components or subsystems) to the user using the Software Acquisition Pathway
Mission Assurance	• The ability of operators to achieve their mission, continue critical processes, and protect people and assets in any operating environment or conditions
Risk	• *Acquisition risk*: Defined as a potential future event or condition that may have a negative effect on achieving program objectives for cost, schedule, and performance; and defined by (1) the probability (greater than or equal to 0, less than 1) of an undesired event or condition and (2) the consequences, impact, or severity of the undesired event, were it to occur • *Mission assurance risk*: Defined as a potential future event or condition that may have a negative effect on mission success

SOURCES:
- *Urgent*: Department of Defense Instruction 5000.02, *Operation of the Adaptive Acquisition Framework*, January 23, 2020; Department of Defense Instruction 5000.81, *Urgent Capability Acquisition*, December 31, 2019.
- *Rapid*: Department of Defense Instruction 5000.02, 2020; Department of Defense Instruction 5000.80, *Operation of the Middle Tier of Acquisition (MTA)*, December 30, 2019; Department of Defense Instruction 5000.87, *Operation of the Software Acquisition Pathway*, October 20, 2020.
- *Mission Assurance*: DoD, *Mission Assurance Strategy*, April 2012; MITRE Corporation, *Systems Engineering Guide*, 2014.
- *Risk*: DoD, *Department of Defense Risk, Issue, and Opportunity Management Guide for Defense Acquisition Programs*, January 2017.

---

[12] Gail Johnson-Roth, *Mission Assurance Guidelines for A–D Mission Risk Classes*, Aerospace Corporation, TOR-2011(8591)-21, June 3, 2011, p. 1. There are 16 key MA processes: design assurance; requirements analysis and validation; parts, materials, and processes; environmental compatibility; reliability engineering; system safety; configuration/change management; integration, testing, and evaluation; risk assessment and management; independent reviews; hardware quality assurance; software assurance; supplier quality assurance; failure review board; corrective/preventative action board; and alerts and information bulletins. MA is also a functional area in acquisition. The contractor and the program office (typically supported by the Aerospace Corporation) conduct these activities to assure mission success. Appendix D provides additional background information on traditional MA practices and standards for space systems. For more detailed information, refer to the Aerospace guidebooks on mission assurance (Johnson-Roth, 2011).

Structure of This Report

Chapter 2 describes the acquisition streamlining pathways and methods used by SSC and Space RCO that we identified through the literature review and interviews with SMEs. Chapter 3 defines risk in the context of SSC and Space RCO programs and describes and bins the potential risks associated with the various streamlining practices, based on the literature review and SME interviews. It also includes potential mitigations and recommendations for these risks, many of which relate specifically to the challenge that if the programs fail to deliver on time, then operational missions could be put at risk. Chapter 4 assesses the potential impact of rapid acquisition strategies on MA specifically. Chapter 5 presents a framework for assessing and managing risks to MA. Finally, Chapter 6 presents our responses to the motivating questions, our conclusions, and recommendations. In addition, several appendixes provide additional details. Appendix A includes a historical snapshot of urgent and rapid acquisition. Appendix B expands on Chapter 2 with additional details on USSF acquisition streamlining practices. Appendix C offers more details on the risks described in Chapter 3. Appendix D provides additional details on the traditional MA processes. Appendix E describes an analysis of MAR data as a potential source of information on acquisition challenges.

2. Acquisition Streamlining Methods

In this chapter, we offer a short overview on acquisition streamlining to provide context. We describe the primary USSF organizations engaging in streamlining and summarize the specific subset of streamlining tools that our research found that the USSF is using to accelerate acquisition.

Background on Acquisition Streamlining

Need for and Approaches to Streamlining

DoD's traditional approach to developing and building weapon and information systems has been criticized for taking too long and costing too much. Multiple process solutions over time have been developed to expedite the acquisition of new capabilities. The most recent policy has resulted in the creation of the AAF via Department of Defense Instruction 5000.02, *Operation of the Adaptive Acquisition Framework*. Figure 2.1 summarizes the rapid pathways that can be used to accelerate acquisition programs.[13]

Program managers (PMs) and the Milestone Decision Authority (MDA) must determine which pathway is most appropriate for their program. One of the main factors determining the most appropriate pathway is the timeline needed for fielding a system. Urgent capabilities usually require that the program not exceed two years, and Middle Tier of Acquisition (MTA) programs are limited to five years each for rapid prototyping or rapid fielding. In addition, recognizing the particular needs of software-intensive programs, the AAF includes a Software Acquisition Pathway to capitalize user engagement and encapsulate the best software development practices to deliver a minimum viable capability release within a year after the date on which funds are first obligated. Our analyses focus on the Urgent Capability Acquisition, MTA, and Software Acquisition Pathways.

---

[13] The AAF has six pathways. We did not include the Major Capability Acquisition (MCA), Defense Business Systems, and Acquisition of Services pathways on this chart because they are not the main pathways that USSF has been utilizing to go faster.

Figure 2.1. New Adaptive Acquisition Framework Includes Three "Rapid" Pathways Allowing Acquisition Professionals to Adopt New or Evolved Procedures

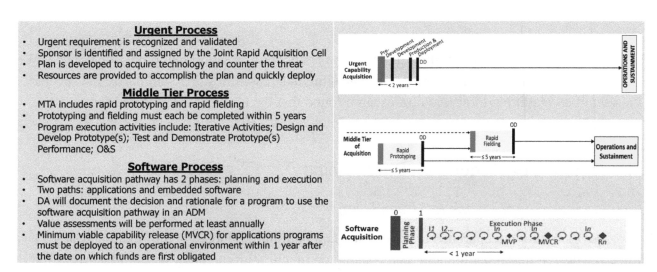

Urgent Process
- Urgent requirement is recognized and validated
- Sponsor is identified and assigned by the Joint Rapid Acquisition Cell
- Plan is developed to acquire technology and counter the threat
- Resources are provided to accomplish the plan and quickly deploy

Middle Tier Process
- MTA includes rapid prototyping and rapid fielding
- Prototyping and fielding must each be completed within 5 years
- Program execution activities include: Iterative Activities; Design and Develop Prototype(s); Test and Demonstrate Prototype(s) Performance; O&S

Software Process
- Software acquisition pathway has 2 phases: planning and execution
- Two paths: applications and embedded software
- DA will document the decision and rationale for a program to use the software acquisition pathway in an ADM
- Value assessments will be performed at least annually
- Minimum viable capability release (MVCR) for applications programs must be deployed to an operational environment within 1 year after the date on which funds are first obligated

SOURCES: Department of Defense Instruction 5000.02, 2020; Department of Defense Instruction 5000.81, 2019; Department of Defense Instruction 5000.80, 2019; Department of Defense Instruction 5000.87, 2020.
NOTES: DA = Decision Authority; ADM = Acquisition Decision Memorandum.

In addition to selecting a pathway, the PM can choose to deviate from traditional acquisition standards in a variety of ways in order to streamline parts of the acquisition process. For example, the PM may decide to use an alternative to the Joint Capabilities Integration and Development System for requirements, as is permitted by law within the MTA Pathway. PMs have also chosen to use Other Transaction Authority (OTA) in contracting, or concurrency of developmental and operational testing. The variety of choices that PMs are able to make often means that no two MTA programs are equal and emphasizes the need to determine the best practices for identifying what fits best for the USSF.

USSF Organizations Are Using Rapid Acquisition Approaches

Given the particular focus on Space RCO and SSC in this report, we summarize relevant characteristics of these organizations to provide context on USSF acquisition (Table 2.1). From an organizational perspective, there are strategic differences between Space RCO and SSC. These differences drive how the organizations are structured. Space RCO is focused on building small teams with a narrow chain of command and on providing highly skilled embedded support to develop and deliver capabilities at the speed of warfighting relevance. SSC is a much larger organization with a mission to "pioneer, develop and deliver sustainable joint space warfighting capabilities to defend the nation and its allies and disrupt adversaries in the contested space domain."[14] SSC builds complex capabilities that require it to communicate with organizations

---

[14] U.S. Space Force, Space Systems Command, "About Space Systems Command," webpage, undated.

within USSF and across the federal government. SSC has numerous integration and synchronization issues that it needs to resolve for space capabilities. As a result, for example, SSC has set up risk boards that meet regularly to assess programmatic risks (see Table 2.1 and Appendix B for further discussion).

Table 2.1. Approaches to Rapid Acquisition and Risk Postures Vary Across the Space Acquisition Community

Space Rapid Capabilities Office	Space Systems Command

• **Mission:** Develop and deliver operationally dominant space capabilities at the speed of warfighting relevance
• **Formed:** 2018
• Mostly uses tailoring of traditional acquisition process for rapid acquisition
  – ACAT I–III programs and urgent needs
  – Program timeline <4 years to fielding
  – Streamlined requirements
  – Use of relatively mature technology
• Lean/agile organization with limited outside dependencies
  – Board of Directors structure for requirements
  – Autonomy of PMs
  – Competencies in-house
  – Integrated team with highly experienced staff
  – Senior leadership support is critical
• Risk-tolerant environment
  – "Block 0"; Transition to traditional program for follow-on
  – Supporting new missions (in direct response to combatant command needs)
  – Shorter design life
  – Highly classified
	• **Mission:** Pioneer, develop and deliver sustainable joint space warfighting capabilities to defend the nation and its allies and disrupt adversaries in the contested space domain.
	• **Formed:** 2021 (from the Space and Missile Systems Center)
	• Utilizing AAF pathways and tailoring of traditional acquisition processes for rapid acquisition
	  – ACAT I–III programs
	  – Program timeline generally 5+ years to fielding
	  – Use of MTA
	  – Other rapid acquisition includes prototyping outside Programs of Record and software pathways
	  – Shortened requirements and contracting process
	  – Fixed schedule
	  – Focused on reducing technology development risk (narrow scope)
	• Different risk postures depending on the mission
	  – Risk-averse for strategic and established missions
	  – Risk-tolerant for new/tactical missions

SOURCES: Shannon Holmes-Terry, Director, NCR Integration Office, Space Rapid Capabilities Office, "Space Rapid Capabilities Office (SpRCO) Overview," Headquarters U.S. Space Force, October 14, 2020; U.S. Space Force, Space Systems Command, undated; discussions with space SMEs.
NOTE: ACAT = Acquisition Category.

What Is the USSF Acquiring Using Rapid Approaches?

Space Systems Command

Within SSC, we identified several programs that are utilizing the MTA and Software Acquisition Pathways.[15] Seven rapid programs were considered in this analysis, six of which are using the MTA Pathway:

---

[15] These programs are widely known and are subjects of a number of U.S. Government Accountability Office (GAO) reports. Refer to the following sources for more detailed information about them: DoD, *Department of Defense Fiscal Year (FY) 2021 Budget Estimates: Air Force Research, Development, Test & Evaluation, Space Force*, February 2020; GAO, *Defense Acquisitions Annual Assessment: Drive to Deliver Capabilities Faster Increases Importance of Program Knowledge and Oversight*, GAO-20-439, June 2020; GAO, *Missile Warning Satellites: Comprehensive Cost and Schedule Information Would Enhance Congressional Oversight*, GAO-21-

- Next-Generation Overhead Persistent Infrared (Next-Gen OPIR) Space

  Next-Gen OPIR consists of five missile warning satellites broken up into two groups: GEO and Polar. This system will replace the legacy program, Space Based Infrared System (SBIRS). Next-Gen OPIR is designed to better face our adversaries and emerging threats.

- Protected Tactical SATCOM (PTS)

  PTS is a part of a larger effort, Protected Anti-Jam Satellite Communications (PATS), and is used to provide anti-jam satellite communications for operators using the Protected Tactical Waveform. The PTS program will deliver two prototype communications payloads to be hosted on U.S. military, commercial, or allied satellites.

- Protected Tactical Enterprise Service (PTES)

  PTES is a ground infrastructure to be used by Wideband Global SATCOM and PTS satellites to provide anti-jam communications for tactical warfighters.

- Evolved Strategic SATCOM (ESS)

  ESS is a follow-on program associated with the strategic mission portion of the Advanced Extremely High Frequency (AEHF) program. It provides worldwide survivable SATCOM to strategic users for nuclear, command, control, and communications. PTS addresses the tactical SATCOM portion of AEHF. The MTA will deliver a prototype payload for the space segment.

- Future Operationally Resilient Ground Evolution (FORGE)

  FORGE is the ground system associated with the Next-Gen OPIR. It consists of command and control (C2), Mission Data Processing (MDP), and Relay Ground Stations (RGS).

- Military Global Positioning System (GPS) User Equipment Increment 2 Miniature Serial Interface (MGUE Inc 2 MSI)

  MGUE Inc 2 is the user equipment program associated with GPS to enable the use of the military signal or the M-code. The program is developing the M-code cards that the military services will integrate into GPS receivers on various platforms. The predecessor program, MGUE Inc 1, developed the M-code cards for ground, air, and maritime platforms. MGUE Inc 2 is developing the M-code cards for munitions, space-based receivers, and handheld receivers. The MSI is one of the MTAs associated with MGUE Inc 2 that will develop the card technology for space-based receivers and munitions.[16] MGUE Inc 2–Hand Held is a second MTA associated with MGUE Inc 2 to deliver modernized handheld receivers.

105249, September 2021c; GAO, *GPS Modernization: DOD Continuing to Develop New Jam-Resistant Capability, but Widespread Use Remains Years Away*, GAO-21145, January 2021a; GAO, *Space Command and Control: Comprehensive Planning and Oversight Could Help DoD Acquire Critical Capabilities and Address Challenges*, GAO-20-146, October 2019.

[16] GAO, 2021a.

One acquisition program with large amounts of software is using practices from the Software Acquisition Pathway:

- Space Command and Control (C2)

  The Space C2 program is a C2 system essential to conducting Space Force operations. It consists of the following product lines to support operational C2 of space forces: space domain awareness; battle management C2 to support the National Space Defense Center; data analytics and visualization; theater/coalition support (to meet Combined Space Operations Center's C2 needs); and DevOps infrastructure.[17]

Space RCO

The Space RCO has classified programs that are mostly tailoring traditional acquisition processes to achieve rapid acquisition. These are generally Acquisition Category (ACAT) I–III programs fulfilling urgent needs with program timelines of less than four years to fielding.[18]

Challenges in Identifying Programs for Analysis

We note here that one of the challenges we faced in conducting the analysis for this project is a lack of existing data in the context of the new USSF, because the AAF and the Space RCO are both so new that the alternative pathways—including MTA—have yet to deliver enough programs to assess outcomes. These outcomes are not yet fully visible and are generally still anecdotal.[19] Likewise, the policy that established the Software Acquisition Pathway was released in October 2020; therefore, most of the software acquisition programs that are implementing this guidance are still in the planning phase. In addition, the Software Acquisition Pathway is based on current commercial best practices in software acquisition, making it impossible to compare outcomes with previous legacy programs.

What Streamlining Techniques Are Used by Space RCO and SSC?

Interviews with Space RCO and SSC staff revealed that the Space Force is increasing its use of urgent acquisition at Space RCO and use of the MTA and Software Acquisition Pathway at SSC. As part of Space RCO's mission, the organization is required to provide "operationally

---

[17] The Space C2 program is also known as Kobayashi Maru, and it is not intended to address all space C2 needs or broader C2 needs (e.g., cross-domain C2). See Space and Missile Systems Center Public Affairs, "Operational Acceptance for Space C2 MINERVA," July 12, 2021; U.S. Government Accountability Office, *Space Command and Control: Opportunities Exist to Enhance Annual Reporting*, GAO-22-104685, December 2021d, p. 10.

[18] According to the Defense Acquisition University:

> Acquisition programs may be assigned an acquisition category. The acquisition category informs the level and amount of review, decision authority, and applicable procedures required for a program. Acquisition category is primarily determined by the expected program cost and/or level of interest. (Defense Acquisition University, "Acquisition Category," Acquipedia, undated)

[19] While we did not include the DAF RCO as part of this analysis, there may be additional lessons learned from how that organization conducted space acquisition in the past.

dominant space capabilities at the speed of warfighting relevance."[20] The discussions also revealed that the organization is focused on using best practices from acquisition of urgent needs over the past 20 years, and that leadership is expending significant effort to adopt the rapid acquisition culture in the service.

Per the Air Force Acquisition Executive's policy guidance, SSC has started utilizing both the MTA and Software Pathways to improve the schedule on programs such as Next-Gen OPIR that would otherwise have been a major defense acquisition program (MDAP) in traditional acquisition. Concerns about the threat from potential adversaries were cited by numerous interviewees as the main driver for the adoption of these practices. We also learned that, currently, speed is the highest priority, and cost is held constant to avoid overruns, which leaves trade-offs to be made in performance. This is a major culture shift in space acquisition, as, traditionally, leadership was opposed to performance trade-offs for schedule or cost.[21] Although we found that these strategies and tactics are being pursued, we did not evaluate their use.

According to interviewees, USSF leadership and culture have been supportive of these streamlining methods. We heard that MTA programs and other rapid programs tend to be prioritized in the functional communities that support the program offices (e.g., contracting, testing, and others), given the importance placed on schedule.

In addition, as emphasized by the interviewees and the literature, the pathways are not unique—they share streamlining methods and offer opportunities to tailor processes—so we focus on streamlining methods here more broadly, rather than linking them to the specific pathways.

Space RCO's Streamlining Tactics

Space RCO is a relatively new organization within the USSF. It is modeled after other successful RCOs. Like those other RCOs, Space RCO uses a lean structure with a short and narrow chain of command, small teams for each program, a small workforce, and embedded functional support. Space RCO is focusing on building a highly skilled and agile workforce for acquisition and support functions.

Streamlining Practices

Many of the streamlining practices that were described in our interviews are consistent with lessons learned and documented in the urgent needs community since 2000. For example, for Space RCO, schedule is the highest priority to counter the adversary's capability. Given the importance of schedule, Space RCO tends to acquire solutions that provide "80 percent" of the

---

[20] Space Rapid Capabilities Office, homepage, undated.

[21] There are examples of where U.S. space systems have traded performance for cost and schedule, but the lack of rapid development causes a self-reinforcing cycle that disincentivizes people from compromising on performance (i.e., if the space community only upgrades every ten years, then the upgrades need to be comprehensive).

needed capability, using a narrow set of requirements. The organization also uses an alternative requirements process (validated by U.S. Space Command [SPACECOM] and assigned by the Board of Directors).

Space RCO generally is not using the formal Urgent Capability Acquisition Pathway; however, there is extensive use of tailoring of traditional acquisition processes. Space RCO programs are different because their more narrow scope and rapid schedule means that the technology readiness level, manufacturing readiness level, maturation, and planning are more readily focused and central to the acquisition effort. In addition, the organization is moving toward applying digital engineering to rapid prototypes and also bringing some testing in-house, while also combining developmental testing (DT) and operational testing (OT) as integrated testing.

Business Practices

Space RCO is using several business-related practices to streamline. For example, Congress granted the Head of Space RCO the authority to use a flexible in-house USSF program element mechanism for highly classified programs to reduce resource approval layers (the SSC does not have such authority). An additional timesaver is that Space RCO has contracting authority in-house: Frequently, there is delegation to contracting officers (within Space RCO) to save time in the contracting process. For contracting mechanisms, Space RCO tends to use Federal Acquisition Regulation (FAR) Part 16.5 (indefinite delivery, indefinite quantity [IDIQ]) and some OTA/multi-award OTA (new) options. Space RCO also employs a business intelligence function to improve industry knowledge. This function allows constant contact with industry, so that Space RCO is kept apprised of the latest technology and companies in the industrial base.

Communication

Finally, given the rapid pace at which Space RCO is acquiring technology, communication with the user and operational communities is critical. A transition cell is in place to bridge acquisition and operational communities. There are also liaisons available to interface with the user community.

Space System Command's Streamlining Tactics

Streamlining Practices

SSC has seven MTA programs (six of the seven are "major" prototyping programs), with the goal of significantly decreasing prior ten-or-more-year acquisition schedules for complex space technology while incorporating agile software acquisition practices. SSC is using the MTA rapid prototyping option for hardware intensive-programs and agile development for software-intensive efforts as a basis for streamlining. The organization has also spent a considerable amount of time streamlining acquisition documentation requirements.

The organization is using alternative requirements processes and a narrow set of requirements or minimum viable product.[22] There is also a focus on using commercial-off-the-shelf (COTS) technology, heritage technology, or lab-transferred technology to reduce development time. Although SSC has some "major" MTA programs,[23] the organization is managing with fewer resources than are traditional MDAPs (i.e., cost is fixed). Despite utilizing fewer resources, the organization is implementing "risk boards" that meet regularly to assess programmatic risks, according to interviewees.

Like the Space RCO, the SSC is trying to identify ways to reduce testing time, which includes conducting development, design, and DT/OT concurrently, as well as employing digital engineering. In addition, the SSC is utilizing multiple iterations of rapid prototyping to "burn down" technical risk, adopting the best practices of digital engineering to rapid prototypes, and using model-based systems engineering tools to generate test requirements documents. These are a few of the new streamlining approaches that PMs are experimenting with as they embrace the new rapid acquisition culture needed in response to the threat.

Business Practices

From a business perspective, SSC has been monitoring and utilizing innovation techniques used in industry. SSC is also utilizing firm fixed price (FFP)/competition in contracting, while working to secure intellectual property rights up front when needed. An example of some current business practices is the enterprise effort to define standards so that commercial and proprietary space payloads can easily integrate with or host military payloads. This effort uses the Program Integration Council (PIC) to ensure alignment with programs that are outside the USSF.[24] The PIC has also been instrumental for programs such as Space C2 that have to integrate with different mission threads, such as space domain awareness and battle management C2.

Summary

Space RCO is a lean and agile organization with limited outside dependencies that operates within a risk-tolerant environment. It was developed to mirror the RCO structure in order to

---

[22] According to Department of Defense Instruction 5000.87, *minimum viable product* is defined as "An early version of the software to deliver or field basic capabilities to users to evaluate and provide feedback on. Insights from MVPs help shape scope, requirements, and design" (Department of Defense Instruction 5000.87, 2020, p. 22).

[23] If an MTA program is expected to require an eventual total expenditure that exceeds the threshold defined pursuant to Section 2302d of Title 10, then the MTA program is a "major system" (Department of Defense Instruction 5000.80, 2019, p. 10). These major systems are at the dollar threshold level of what have traditionally been referred to as *major defense acquisition systems*. According to Section 2302d of Title 10: (1) the total expenditures for research, development, test, and evaluation for the system are estimated to be more than $115 million (based on fiscal year 1990 constant dollars) or (2) the eventual total expenditure for procurement for the system is estimated to be more than $540 million (based on fiscal year 1990 constant dollars).

[24] Sandra Erwin, "Space Force, DoD Agencies, NRO Try to Get on the Same Page on Future Acquisitions," *Space News*, September 22, 2020.

produce capabilities quickly to keep pace with the threat. Space RCO is using mostly tailoring of traditional acquisition process for rapid acquisition. Space RCO is attempting to have program timelines under four years.

SSC is much larger than Space RCO. It conducts the acquisition of more complex capabilities that require significant integration and synchronization. More recently, it has been using the AAF pathways and tailoring of traditional acquisition processes for rapid acquisition of the next generation of space capabilities. Program timelines are generally five or more years to fielding (except for capabilities in the MTA and Software Acquisition Pathways).

Acquisition is a complex process involving many steps. Text Box 2.1 offers a taxonomy of government functions related to acquisition, many of which offer opportunities for streamlining to accelerate acquisition. We include this information for background context, as many of the risks we identify relating to rapid acquisition go beyond the specific functions over which the PM has direct control. Interviewees provided us examples of how they are using rapid acquisition in some of these acquisition functions. Our analysis is not a comprehensive list of everything that is being done within each of these acquisition functions. Interviewees did not contribute information to all of these functions. In addition, as much as possible, we tried to follow this taxonomy when presenting the various practices identified by interviewees throughout multiple tables in Chapter 2, Appendix B, and Appendix C.

Text Box 2.1. Taxonomy of Government Functions Related to Acquisition

1. Program Management/Manager
   1.1 Business case and economic analysis
   1.2 Affordability analysis
   1.3 Acquisition strategy
   1.4 Risk management
   1.5 Technical maturity
   1.6 Personnel and team management
   1.7 Business and marketing practices
   1.8 Configuration management
2. Research and Development (R&D)
3. Engineering
   3.1 Systems engineering
   3.2 Facilities engineering
   3.3 Software/IT
4. Intelligence & Security
   4.1 Cybersecurity
   4.2 Program Protection
5. Test and Evaluation (T&E)
   5.1 Developmental T&E
   5.2 Operational T&E
6. Production, Quality, and Manufacturing (PQM)
7. System and Operational Issues
   7.1 Spectrum (frequency allocation, emissions, etc.)
   7.2 Environmental
   7.3 Energy
8. Product Support, Logistics, and Sustainment
9. Financial Management

10. Cost Estimating
11. Auditing
12. Contract Administration
    12.1 Contracting actions
    12.2 Contracting strategy
    12.3 Contract peer review
    12.4 Acceptance of deliverables
13. Purchasing
14. Industrial Base and Supply-Chain Management
15. Infrastructure and Property Management
16. Manpower Planning and Human Systems Integration
17. Training and education
    17.1 Training and education for government execution
    17.2 Training and education for acquired systems
18. Disposal

Acquisition Interface Functions

19. Requirements: receive, inform, and fulfill
20. Acquisition Intelligence: request, receive, and respond
21. Legal Counsel: request and act upon

SOURCE: Philip S. Anton, Megan McKernan, Ken Munson, James G. Kallimani, Alexis Levedahl, Irv Blickstein, Jeffrey A. Drezner, and Sydne Newberry, *Assessing Department of Defense Use of Data Analytics and Enabling Data Management to Improve Acquisition Outcomes*, Santa Monica, Calif.: RAND Corporation, RR-3136-OSD, 2019.

Our discussions with USSF acquisition professionals revealed that they are using the AAF generally and many streamlining techniques specifically to accelerate their programs, including tailoring. Table 2.2 provides examples of streamlining tactics mentioned in interviews with Space RCO and SSC staff. For example, programs described implementing processes to reduce the time needed to identify contractors and to get them on contract, shortening the requirements process, reducing the numbers of layers of bureaucracy required for approvals, and reducing needed documentation. They are also generally starting with more mature technologies (hence reducing some technical risk) and narrowing development scope and requirements. Lastly, they aim to deliver capabilities incrementally, rather than going after the full suite of capabilities in a single program. This last point is of particular importance because the space community has had challenges in the past in incrementally delivering capabilities. Specific examples of challenges in delivering capabilities incrementally could include the transitions from Defense Meteorological Satellite Program (DMSP) to National Polar-orbiting Operational Environmental Satellite

System (NPOESS) (and now Weather System Follow-on Microwave [WSF-M]) or Space Based-Infrared System (SBIRS) to Next-Generation OPIR System.

Chapter 3 will review the risks that USSF needs to consider as it fully embraces streamlining in acquisition.

Table 2.2. Summary of Accelerated Acquisition Strategies and Tactics Identified by Interviewees

Acquisition Processes and Functions	Space Rapid Capabilities Office (Space RCO)	Space Systems Command (SSC)
Acquisition process: pre- or post-Adaptive Acquisition Framework (AAF)	• Uses tailoring of traditional acquisition processes extensively • Uses urgent capabilities best practices extensively	• Uses Middle Tier of Acquisition (MTA) (rapid prototyping for hardware intensive) and agile development for software-intensive efforts • Streamlines documentation requirements
Requirements	• Defines "80 percent" end-to-end capability using narrow set of requirements • Uses alternative requirements process (validated by United States Space Command [SPACECOM] and assigned by the Board of Directors)	• Uses narrow set of requirements • Uses alternative requirements process (in context of Middle Tier of Acquisition prototyping) • For software, defines minimal viable product that be delivered
Resources	• Uses Head of Space RCO spending authority to reduce resource approval layers	• Manages with fewer resources than traditional Major Defense Acquisition Programs (MDAPs) (i.e., cost is fixed)
Research and development	• Uses commercial-off-the-shelf (COTS) and heritage technology to reduce development time • Acquires systems with narrow research and development requirements • Monitors and uses innovation in industry	• Uses COTS, heritage technology, or lab-transferred to reduce development time • Monitors and uses innovation in industry
Engineering	• Applies digital engineering to rapid prototypes • Uses multiple iterations with 80 percent solution	• Applies digital engineering to rapid prototypes • Uses rapid prototyping and multiple iterations to burn down technical risk • Uses model-based systems engineering tools to generate test requirements documents
Test and evaluation	• Brings some testing in-house/combined developmental testing (DT)/operational testing (OT) integrated testing	• Conducts development, design, and DT/OT concurrently
Contract administration	• Employs business intelligence function to improve industry knowledge • Uses FAR Part 16.5 (Indefinite delivery, indefinite quantity [IDIQ]); some Other Transaction Authority (OTA)/multi-award OTA (new) • Has contracting authority and delegates to contracting officers (within Space RCO)	• Uses Firm Fixed Price (FFP)/competition in contracting • Uses OTAs and IDIQ during MTA • Uses Space Enterprise Consortium (9+ companies)

Acquisition Processes and Functions	Space Rapid Capabilities Office (Space RCO)	Space Systems Command (SSC)
Training and education	• Focuses on highly skilled/agile workforce for acquisition and support functions	• Needs to ensure good training programs because there is large turnover in software workforce (and acquisition generally due to additional factors)
Integration and synchronization	• Focuses on building end-to-end capability	• Uses an enterprise effort to define standards so that commercial and proprietary space payloads can easier integrate with or host military payloads; Program Integration Council (PIC) • Creation of the Portfolio Architect (SSC/ZA) function under SMC 2.0 and/or the realignment with a Space Systems Integration Office under the current commander
Transition to fielding/sustainment	• Transition cell in place to bridge acquisition and operational communities • Liaisons available to interface with the user community	• Secures intellectual property rights up front as needed
Organizational structure and culture	• Schedule is the highest priority to counter adversary's capability • Uses a lean structure with short/narrow chain of command, small teams for each program, small workforce, and embedded functional support	• Has risk boards that meet regularly to assess programmatic risks

SOURCE: RAND discussions with SMEs.

3. Identification of Key Risks in USSF Rapid Acquisition

This chapter describes key risks from current USSF rapid acquisition efforts that were identified during this analysis along with a brief introduction on risk to orient the audience. (A more comprehensive list and discussion of those risks along with a discussion on methodology can be found in Appendix C.) During our interviews with space experts, we requested information on current risks that each person was monitoring as related to USSF rapid acquisition. The SMEs responded with information that was based on their current position and on prior experience. The interviewees offered insight into a significant number of risks that were actively being monitored or were something that might be on the radar in the future; however, not all can be weighed as critical for leadership attention. This chapter provides the main risks identified through this analysis that leadership needs to monitor across all rapid acquisition efforts in USSF. In addition, we provide mitigations if they are available. These derive from interviews, the literature, and team analysis.

Risk in Defense Acquisition

One of the objectives of the Defense Acquisition System (DAS) is to "Deliver Performance at the Speed of Relevance."[25] DoDD 5000.01 provides some overarching ways that this can be accomplished:

- Empower PMs.
- Simplify acquisition policy.
- Employ tailored acquisition approaches.
- Conduct data-driven analysis.
- *Actively manage risk.* [emphasis added]
- Emphasize product support and sustainment.
- Use an adaptive acquisition framework to emphasize these principles.[26]

"Actively managing risk" is a key tool for the acquisition enterprise. It is arguably even more critical in the current rapid acquisition environment where speed is the highest priority, cost is fixed, and trade-offs are being made with performance. It is very difficult to predict what might happen as USSF programs are moving at a faster pace than in the past, so risk management is critical.

Managing risk is the focus of a significant set of DoD policies, including DoD's *Risk, Issue, and Opportunity Management Guide for Defense Acquisition Programs,* which provides a DoD

---

[25] In the Chief's Planning Guidance, one of the CSO's stated priorities is to "deliver new capabilities at operationally relevant speeds" which aligns with this goal (1st Chief of Space Operations, 2020).

[26] Department of Defense Directive 5000.01, 2020, p. 4.

Risk Management Framework. Figure 3.1 provides DoD's framework for how the workforce can think through potential events that might happen and potential consequences of those events.

Figure 3.1. Overview of Potential Sources of Program Risks, Issues, and Opportunities

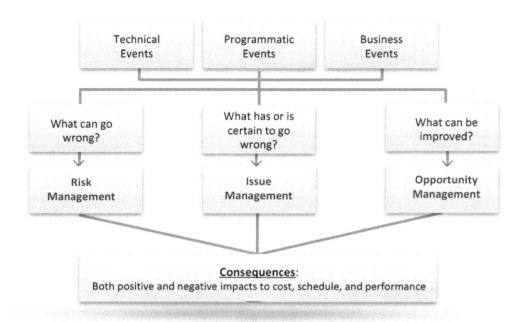

SOURCE: DoD, 2017, p. 3.

DoD's Risk Management Framework explains that *technical*, *programmatic*, and *business* events may lead to risks, issues, or opportunities, each with cost, schedule, or performance consequences. Table 3.1 defines these categories in more detail.

Table 3.1. Technical, Programmatic, and Business Events Defined

Technical	• Risks that may prevent the end item from performing as intended or from meeting performance expectations • Can be internally or externally generated • Typically emanate from areas such as requirements, technology, engineering, integration, test, manufacturing, quality, logistics, system security/cybersecurity, and training
Programmatic	• Nontechnical risks that are generally within the control or influence of the PM or Program Executive Office (PEO) • Can be associated with program estimating (including cost estimates, schedule estimates, staffing estimates, facility estimates, etc.), program planning, program execution, communications, and contract structure
Business	• Nontechnical risks that generally originate outside the program office or are not within the control or influence of the PM • As appropriate, business risks should be escalated up the chain to the appropriate level • Can come from areas such as program dependencies; resources (funding, schedule delivery requirements, people, facilities, suppliers, tools, etc.); priorities; regulations; stakeholders (user community, acquisition officials, etc.); market factors; and weather

SOURCE: DoD, 2017, pp. 22, 79.

Within the USSF acquisition community, this framework is used by the PMs, program executive offices (PEOs), and functional communities (testing, requirements, etc.). It provides a common way to discuss and think about risk in reference to acquisition. It is also required to be used by the PMs and PEOs as they regularly report program status to the Air Force Acquisition Executive through MARs. We aligned our analysis to the DoD framework and used the same terminology when trying to bin the risks (see Appendix C).

Key Risks and Potential Mitigations Identified in Discussions

During our interviews, SMEs shared a long list of risks related to USSF pursuing acquisition streamlining, most typically focused on the multiple risks that they identified as related to their current role in the organization (PM, PEO, contracting, testing, engineering, operator, etc.). Given the broad knowledge they acquired over their long careers as military, civilians, or contractors, they often also offered some additional insights into potential risks.

Although we asked specifically about risks related to acquisition streamlining, the risks that were raised were typically risks found more generally in acquisition, but the interviewees noted that the urgency of rapid acquisition enhanced the criticality of some risks. For example, all acquisition programs face risk related to resourcing, but given the tight timelines for rapid programs, movement of money out of a program can be devastating to the program schedule and hence to delivery of a needed capability.

Interviewees also noted that there are potential risks to any program and that it is not possible to foresee everything that could go wrong. This is still true when programs make additional trade-offs or accept more risk to maintain a rapid schedule.

21

The interviewees noted that the workforce typically has a good handle on *technical-related risks* associated with the program. The space community has focused on complex technical problems for decades and has, likewise, spent a considerable time building in processes to account for risk both through the contractor's processes and through the government's processes. Additionally, many rapid programs are using more mature technologies, such as COTS or technologies from past programs, to reduce technical risks. That said, some SMEs indicated that technical issues are usually uncovered during integration and tests, and the current rapid programs have not yet reached that point.

Programmatic-related risks (e.g., cost estimating, contracting) also seem to be manageable by program offices because of the existence of established and well-understood practices. Interviewees offered fewer concerns about risks in this area. For example, in the case of contracting, the workforce has spent the past 20 years procuring urgent capabilities, which has provided a lot of best practices in this area for current rapid programs to follow. That said, execution challenges sometimes arise relating to tighter timelines.

Finally, *business-related risks* were the area where the SMEs focused a lot of their attention when discussing risks. These are typically nontechnical events that generally originate outside the program office and include, but are not limited to, resources, organizational priorities, regulations, and stakeholders (e.g., user community).

The key risks that we identified and describe below are for the most part in no particular order because, in the absence of delivered programs, we are unable to estimate which would have the greatest impact. These key risks all have a potential impact on MA, because if they result in a necessary capability not being delivered in a timely way, then the mission is at risk. We close the chapter by describing a risk specifically related to the USSF's management of MA.

Finally, if we were able to identify any potential solutions based on SME discussions, the literature, and our own acquisition subject-matter expertise, we include a description of those potential mitigations.

Key Risk 1: If the space operational community is not closely aligned with the space acquisition community, then the utility of capabilities being transferred may be diminished or delayed

DoD's acquisition community has received a clear mandate and tools to move faster by Congress and DoD leadership. There are fundamental changes happening in the acquisition community as the AAF takes hold. One predicted outcome is the movement of more capability from the acquisition to operational communities (within a shorter period of time). This leads to a series of questions that arose in discussion that the USSF operational community needs to consider in the near term:

- Has the operational community changed any processes as the acquisition community has begun to go faster?

- Will the operational community be ready to catch the wave of capabilities from Space RCO?
- How is the operational community thinking through operational acceptance (i.e., the transfer of a capability from the acquisition to the operational community)?
- Do space operators have time to get involved early enough in the requirements development, transmission, and subsequent acquisition processes, given that they have "day jobs"?
- Are operator facilities prepared to accept new technology (e.g., do they have sufficient power, air conditioning, proper accreditation)?
- Will users have sufficient training to use new high-tech capabilities?
- Has the acquisition community adapted its delivery of supporting materials to help the operational community digest the new capability?

We heard some clear concerns from both the operators and acquisition personnel related to this risk. There is a concern that the user community may not have taken sufficient steps to align with the new acquisition pace (i.e., revamped processes).[27] The operational community may also be expecting the same level of capability and training documentation as in the past, but we already know that performance is the main trade-off to schedule maintenance. Likewise, the operational community does not have a significant amount of time to get new capabilities up and running and so must prepare to have sufficient operators to accept large number of capabilities in a compressed timeline. In addition, we heard from SMEs that a clearer priority list of capabilities from leaders of the operational units accepting the new capabilities would be useful for the acquisition community to ensure that the operators are getting what they need when they need it and that the operational community has the resources to support what is being sent over.

Recommended Mitigations

Two fundamental shifts must continue to occur within the operational community: The operational community needs to be involved earlier and more often in rapid acquisition (e.g., MTAs, software, etc.), and changes may be needed on the operational side to accept more capability faster and with less formalized training materials. While these shifts are necessary for all acquisitions, rapid acquisition requires that these shifts now receive more urgency and priority in the operational community. An overall list of priorities from leadership will also help both the acquisition and operational communities prioritize what is coming through the acquisition pipeline. Finally, as suggested in a prior report, "dissolving seams traditionally separating operators and acquirers so that all understand both technology and operations; operators will know how technology flows and changes, and acquirers will know how technology is implemented."[28]

---

[27] Additional analysis is needed to document the exact operational communities that are experiencing these challenges.

[28] Shelton et al., 2021, p. 61.

The USSF could also explore existing coordination tools. For example, the MITRE Corporation created a risk management software tool called the Risk Matrix to help "identify, prioritize, and manage key risks on a program."[29] The goal of this tool is to "capture identified risks, estimate their probability of occurrence and impact, and rank the risks based on this information."[30] This or other tools could be used by PMs to understand ahead of time if risk will bleed over to operators. One potential use for this tool on rapid acquisition programs is to identify potential risks for transferring the capability to the operational community.

Key Risk 2: If resourcing is not aligned with rapid acquisition program schedules, then leadership will need to make tough trade-offs resulting in degraded operational performance to maintain accelerated schedules

During our discussions, various facets of resourcing were called out by interviewees as problematic. Although the challenges regarding resources are not new in the acquisition community, these resource challenges are ways that risk is being elevated in rapid space programs. PMs are encountering significant programmatic business risks, as well as obstacles already present in congressional and DoD authorizations. Some of these challenges include the following:

- Programs that require funding to "ramp up" at the beginning do not have access to the funding needed.
- Funding is not available on the program's required schedule (i.e., funding is unreliable or poorly timed).
- Funding is removed from the program during the compressed schedule.
- PMs need to use a Program Objective Memorandum (POM) for acquisition, operations, and maintenance, but the traditional three-year POM cycle is too long for the compressed rapid acquisition schedule, which is under five years. This affects software in particular because a delivery is required within the first year.
- Traditional budget models for funding programs do not align with the long lead time needed for funding by industry for deliveries.
- Congress may deny above-threshold programming requests to cover urgent needs.

Recommended Mitigations

While some of the above challenges can be addressed within the USSF resource community, others must be addressed by the larger DoD resource community and Congress. For example, the misalignment of the POM cycle with the rapid acquisition programs that we identified in this analysis needs to be addressed by Congress. In addition, it may or may not be appropriate to advocate to Congress for funds to be available up front for MTA or Urgent Capability

---

[29] MITRE Corporation, "MITRE Systems Engineering Guide: Risk Management Tools," webpage, undated.

[30] Pamela A. Engert and Zachary F. Lansdowne, *Risk Matrix User's Guide*, Version 2.2, MITRE Corporation, November 1999.

Acquisition Pathway segments of larger programs to get the right resources in place on time. As the USSF moves to using incremental capability acquisition more often for large, complex programs that typically correlate with large costs and long schedules, compressing their schedules requires an ever larger "up-front" cost. Subdividing the acquisition of a capability is a means to address this problem and is explicitly accounted for in the minimum viable product approach of Software Acquisition. Traditional budget models need to be revised to reflect the new reality of rapid acquisition to ensure that programs have the funding up front for the long lead time required by industry.

Managing space capabilities as a portfolio rather than as individual acquisition programs allows better allocation of resources to higher priority activities and will help ensure that capabilities are delivered according to accelerated schedules.[31] In addition, the programs may be able to shift the delivery of a capability to a later date if resources are not available, or provide minimal capability up front to match resources.

Key Risk 3: If the USSF is conducting threat-based acquisition, then the shortage of on-site intelligence and cybersecurity personnel within the rapid acquisition life cycle will negatively affect this mission

During our discussions, we heard that there are several issues involving intelligence and cybersecurity that may increase risk in rapid acquisition. First, and most important, there appears to be a severe shortage of personnel on-site in USSF in these two key areas; however, the USSF is relying on intelligence to tailor what it is acquiring and cybersecurity-testing, so that adversaries cannot destroy the capabilities or that recovery is possible. These gaps must be closed in order for acquisition to be threat-based. In addition, the nature of the intelligence needed is changing. Intelligence needed to make these acquisition decisions is dynamic and needs to be infused in programs rapidly. This includes intelligence collected from the operational community. However, the acquisition community may not have necessary clearances to see all the intelligence needed to make critical acquisition decisions. The intelligence community similarly faces limitations on access to information about highly classified programs, and thus may not target their information gathering appropriately for these programs.

In addition to the personnel shortage, time and/or facilities appear to be insufficient to fully test for cybersecurity considerations in an "operational" environment. Likewise, supply chain security may be difficult to achieve during compressed timelines.

Recommended Mitigations

We realize that these are difficult challenges for USSF leadership to address. The labor shortages for intelligence and cyber are pervasive across DoD. Therefore, it will be a challenge to fill the gap without figuring out ways to use special hiring practices or finding other ways to

---

[31] Shelton et al., 2021, p. 19.

attract talent. However, there may be ways of better using existing resources—for example, by aligning intelligence professionals by PEO rather than in a central intelligence function, so they can develop a good sense of the acquisition programs that they can inform. The issues of a lack of insight into intelligence information on the acquisition side, and a lack of insight into special access programs on the intelligence side could be addressed by extracting less-classified insight from higher-classification documents for wider dissemination and investing in improving appropriate clearances. Another option is to reduce the barriers with the operational community so that knowledge of the current threat deriving from that operational context can inform rapid acquisition programs. Finally, this is another area where a "bite-sized," incremental acquisition approach can help. The shorter schedules and smaller capability increments mitigate the risks associated with span of intelligence needed and its currency/dynamic attributes.

The issues of insufficient facilities to test for cybersecurity may need to be considered as the USSF is transforming its testing processes and facilities to accommodate the rapid changes in acquisition. In addition, ensuring that cyber resiliency and safety requirements are addressed early in contracts and test protocols, respectively, may help mitigate cybersecurity challenges.

The USSF already spends a lot of time meeting regularly with peers across industry to gain up-to-date knowledge of the challenges faced by contractors, suppliers, and the industrial base in order to address supply chain security concerns. Another mitigation for issues regarding supply chain security might be to leverage traditional programs that have the resources and the time for due diligence. Although rapid programs might or might not be able to share parts with these programs, they can obtain information about vendors who use good practices and any problematic vendors to avoid.

Key Risk 4: If the USSF's testing community is not involved in rapid programs at the appropriate times and levels, serious schedule challenges may result

During the discussions, SMEs often noted the critical function that testing plays in acquisition and shared their observations that this function is not always prioritized and is frequently underresourced. Furthermore, it is not always clear how much testing is needed to understand whether or not the capability will work and whether or not its addition to the overall space architecture might cause problems elsewhere. In addition, the USSF's move to threat-based acquisition may require new and different test facilities compared with those used in the past and, potentially, additional testers to keep up with the workload.

SMEs also discussed some major changes happening in the space testing community. Some organizational movement is occurring—for example, Air Force Operational Test and Evaluation Center (AFOTEC) is moving to Space Training and Readiness Command (STARCOM), some testing is being brought within Space RCO, and integrated test forces are being stood up that will help combine DT and OT. The test community also faces challenges with trying to move to continuous testing under development, security, and operations (DevSecOps) in software-heavy programs (i.e., "build as you fly").

Despite the above churn, there are rapid programs that currently need testing because PMs need to prove to senior leadership that technology will be ready for operational use during the allowable schedule (i.e., PMs need to test to prove that the capability can address the threat and that the operators must accept any identified risks). If testing is not prioritized, then senior leadership may not have confidence that the technology will be operationally effective before approving it.

Recommended Mitigations

The program office, in collaboration with the test community and the operational community, could develop an alternative testing strategy that is better postured to address the many uncertainties about the threat environment (i.e., there has never been a conflict in space before, and potential threats and their capabilities are rapidly changing). Because of these uncertainties, a perfect test solution may not be realistic in the near term, and a higher risk tolerance may be necessary. Recognizing these constraints, the test objectives may need to aim to *reduce the uncertainties* about the system's operational effectiveness, rather than *demonstrate* the system's operational effectiveness. Testing approaches such as the use of modeling and simulation or ground-based tests could prove useful in reducing the uncertainties about the system's operational effectiveness, as well as help operators understand what the system's range of capability is to be able to respond to adversary actions with all the tools at their disposal. We recognize that this testing challenge affects all rapid programs (and possibly traditional space programs) as the USSF builds the necessary knowledge base and foundational infrastructure (e.g., space test range, Joint Munitions Effectiveness Manual for space). Thus, further explorations of alternative testing strategies for the space warfighting environment may be warranted.

USSF is already in the process of employing some streamlining practices involving testing:

- conducting testing in parallel with other parts of acquisition process
- using an independent third-party to determine that the software code is usable
- meeting with testing representatives early (including the Director, Operation Test and Evaluation [DOT&E]) while planning rapid programs.

These changes will help alleviate some of the testing burden, given the rapid schedules. However, given the organizational changes and potential pipeline of rapidly acquired capabilities, leadership may want to continue to monitor whether the test facilities and personnel are adequate for keeping up with rapid acquisition.

Key Risk 5: *If the USSF does not fully recognize and plan for challenges related to employing modern software practices, then the organization will likely experience negative effects during legacy, current, and future program integration*

DoD is moving toward modern software practices and has designated a specific pathway in the AAF for software-intensive systems. Under heavy congressional scrutiny, DoD conducted several pilot programs and has gathered lessons learned to help the acquisition community make the transition. During our discussions, the USSF SMEs talked frequently about the movement from other methods of software development (e.g., waterfall, in which development is divided into sequential phases with limited overlap) to modern practices, including the following, described in the Software Acquisition Pathway guidance:

> This pathway integrates modern software development practice[s] such as Agile Software Development, DevSecOps, and Lean Practices. Capitalizing on active user engagement and leveraging enterprise services, working software is rapidly and iteratively delivered to meet the highest priority user needs. Tightly coupled mission-focused government-industry software teams leverage automated tools for development, integration, testing and certification to iteratively deploy software capabilities to the operational environment.[32]

USSF SMEs expressed several concerns regarding this movement in software development. First, there are challenges trying to integrate new software using these practices with software developed under other methods. Second, the organization is using these practices in an uneven manner (i.e., the space acquisition community is using a mix of old and new practices). Third, some contractors have moved to these newer practices, while others have not. Fourth, the USSF lacks sufficient personnel with the appropriate skills for this type of software development. We also heard that workforce training is an important requirement for moving USSF forward using these new software practices. Finally, integrating the space segment that is being delivered in a traditional way with the ground segment where developers are employing software drops to align with DevSecOps is one of the key areas that needs improvement. The ways in which software developers are measuring progress and in which hardware developers are doing the same are currently incompatible for seamless integration between the space and ground segments.

Recommended Mitigations

There are some obvious mitigations to the challenges described above. Training is needed in these new software development practices for the USSF workforce. The Defense Acquisition University has provided abundant materials on the new Software Acquisition Pathway, and it offers training classes.[33] The Office of the Secretary of Defense has a mentor program, which was established as part of the pilot program required by Congress. This mentor program helps give the services insight into appropriate use of the Software Acquisition Pathway (e.g., setting up testing, contracting, a user agreement).

The integration of software developed under older and newer practices may be a greater challenge. This is a technical challenge that needs to be monitored to avoid affecting the delivery schedules of newer software.

---

[32] Defense Acquisition University, "Software Acquisition," webpage, undated.

[33] Defense Acquisition University, "Software Acquisition," webpage, undated.

Key Risk 6: If the USSF does not take sufficient time to plan for integration early in the rapid acquisition life cycle, then integration may be a significant challenge that will delay transition

Integration and synchronization of space capabilities are challenges that have long led to schedule delays. Our discussions identified a list of potential integration and transition risks associated with rapid acquisition. For integration and transition, we identified the following challenges. Some are challenges that all acquisition faces, but the schedule pressures exacerbate these issues:

- In rapid acquisition, multiple interdependent rapid programs are likely needed to deliver end-to-end capabilities, which complicates integration.
- Asynchronization of funding and delivery schedules of interdependent programs is a potential challenge.
- Complexity in coordinating and synchronizing with interdependent systems as they each mature and evolve rapidly will also be a challenge.
- Given that integration is usually later in the acquisition process, resources for integration may be limited.
- SSC's limited integration experience may also complicate integration. For example, the government may need to be the integrator for MTA programs, but the government has not taken on this role consistently and has had challenges as integrator in the past.
- An integration plan appears to be lacking for many interdependent programs.
- Alignment and integration across the segments (e.g., space, ground, space *and* ground) is managed by different contractors, which may lead to synchronization problems despite the compressed schedule.
- Operators may not be ready to operate and sustain the capability if they are not prepared to receive these capabilities on a compressed schedule.
- For rapid prototyping, transitioning to the Major Capability Acquisition Pathway would require documentation and funding that could lead to delayed fielding (although significant tailoring is still possible).

Recommended Mitigations

We have provided a long list of potential challenges that involve integration and transitioning rapid acquisition capabilities. We propose some potential mitigations for various aspects of these challenges. First, the USSF can focus on modularity and owning the technical baseline (especially the interfaces) to overcome a lack of standards and to reduce the government's dependence on a vendor's proprietary work.[34] Second, the USSF needs to continue to support transition mechanisms in place (e.g., Space RCO's Operations Transition Cell) to ensure that personnel and facilities are available in time for new capability delivery. Within the operational community, there also needs to be a liaison with the acquisition community to help operators prepare for the new capability delivery. Finally, integration cannot be an afterthought:

---

[34] See Will Roper, *There Is No Spoon: The New Digital Acquisition Reality*, October 7, 2020.

Integration and transition issues should be identified early in the rapid acquisition efforts and appropriately addressed and resourced.

Key Risk 7: If the USSF is fully utilizing alternative requirements processes for determining rapid acquisition program requirements, then the user might not get the right capability if those processes are not performing as expected

Our discussions revealed a variety of positions regarding the use of alternative requirements processes during rapid acquisition. The discussion revolved around the alternative requirements process permitted by law under the MTA Pathway and also the requirements validated by SPACECOM and the Board of Directors for Space RCO. The main benefits were time savings from the lengthy Joint Capability Integration and Development System process and the ability to bake in requirements flexibility. However, we also heard the reverse: that requirements are not sufficiently narrow to acquire the product within the mandated schedule. In addition, there may not be sufficient enterprise-wide coordination on prototype requirements.[35]

Recommended Mitigations

The idea of defining an "80 percent" solution using a narrow set of requirements has proven successful in acquisition outcomes in urgent needs over the past 20 years. The USSF is following this best practice. We also heard that the USSF is incorporating real-time operational feedback to generate requirements and is considering reducing requirements for later increments to lower risk. The organization may want to review periodically whether the set of requirements changed dramatically from the start to the end of the rapid acquisition. PMs told us that they are making trade-offs in performance—but not in schedule or cost—in rapid acquisition. This newer way of making trade-offs in the space acquisition community deserves scrutiny to ensure that the alternative requirements processes are helping produce positive outcomes. Finally, incremental capability acquisition may help mitigate the risk of insufficiently narrow requirements by breaking the acquisitions into smaller pieces.

Key Risk 8: If the USSF continues to conduct mission assurance as has been done in the past, this may also hamper accelerated schedules in rapid programs

MA, or the ability of operators to achieve their mission, is another critical component that needs to be considered by both the acquisition and operational communities. In our discussions, we heard that MA requires consistent planning and communication among these communities. We also heard that the space community has a long history of solid MA and established methodology and processes. However, several questions were raised when considering rapid acquisition and MA:

---

[35] This analysis did not address joint requirements, so we do not know how often or whether the USSF is using an alternative requirements process for joint requirements.

- What is acceptable MA in rapid acquisition?
- If the systems are no longer exquisite solutions, how should MA change?
- What changes to MA should be considered up front and addressed through risk mitigation?
- How are safety standards, testing, and cyber modified in a rapid acquisition environment?

During our discussions, several challenges were raised that were specific to MA. The first is that "MA" means a lot of different things depending on where people sit in the acquisition and operational communities. This lack of standard understanding across these communities may cause communication challenges. In addition, with the accelerated programs, time or resources may be insufficient to do traditional MA, and tailoring MA for rapid programs may create tensions in the traditional MA model of staffing and resourcing that need to be considered. The acquisition community may also feel that schedule pressures force less rigor in MA, which would then transfer MA risks to the user or the follow-on program.

Recommended Mitigations

The challenges above are worth examining further. We developed an analytic framework to assist the USSF in identifying and overcoming these potential MA challenges. We present this model in Chapter 5. Rapid acquisition focuses on improving the probability of success of a mission rather than elimination of risk to that mission. It is about getting warfighters (even partial) solutions sooner. This means a truly mission-focused MA versus a perfect system-focused MA.

In addition, it is worth noting that there is potential for the new, rapidly acquired capabilities to improve MA. Rapid fielding should increase MA even though new capabilities come with other performance risks. Also, an MA focus on mission means that, in some cases, a failure of a new capability may be better (or at least not worse) than not having the capability at all. This has been documented in lessons learned from urgent capabilities acquisition over the past 20 years.

Summary

This chapter identified key risks that were most frequently cited throughout the interviews and in the literature and summarizes suggested mitigations. Some of these risks are common to both traditional and rapid acquisition, but introduction of shorter timelines for delivery of capabilities makes the risks more salient. These risks include

- lack of alignment of the acquisition and operations communities
- unreliable or inadequately timed resources
- a shortage of on-site personnel with expertise in cyber and intelligence
- a lag in development of needed test capabilities and infrastructure
- challenges in aligning legacy and modern software development practices
- failure to consider and plan for systems integration
- alternate requirements processes that may not yield needed capabilities.

Each of these risks can, itself, threaten MA, but the final and probably overarching concern is that traditional MA processes may not prioritize the accelerated schedules of rapid programs.

To complement the findings from the interviews and literature review discussed in both Chapters 2 and 3, we conducted an analysis on some MARs, which included a collection of issues identified by PMs of MTA programs. The data were aligned with DoD's Risk Assessment Framework described in this chapter. The data presented some challenges, including minimum available data in the MARs risk section, so we searched for risks associated with the programs throughout the full text of each MAR. The analysis of this information is provided in Appendix E. The MARs represent a potentially useful source of data for leadership on risks and could be used to compare streamlined and traditional programs to determine whether they perceive different risks in their programs.

4. Mission Assurance in Rapid Acquisition

The central research question in this analysis aims to characterize how streamlining techniques used in rapid acquisition affect MA. Characterizing a causal relationship between streamlining techniques and MA (the outcome) is very complex. Many factors unrelated to streamlining techniques (e.g., unplanned budget cuts, inability to hire the right experts) could affect MA, and it is difficult to isolate the impact of those factors from the impact of the streamlining techniques. Further, risks evolve over the program life cycle in a dynamic, nonlinear manner. There are controls in place (e.g., design reviews or tests, which are components of the MA process) to identify and address risks, potentially preventing the risk source from manifesting into an adverse outcome that degrades MA. As a result, it is difficult to identify a traceable path from a specific streamlining method to a specific outcome as it relates to MA.

That said, it is important to understand how rapid acquisition impacts MA more broadly to inform how risks to MA can be managed appropriately for rapid acquisition. To that end, we (1) examined how MA objectives and MA approaches differ for rapid acquisition compared to the traditional acquisition and (2) identified potential challenges that rapid acquisition might introduce in achieving MA. These issues have implications for how MA risks are managed for rapid acquisition. Thus, we developed a framework for assessing MA risks from rapid acquisition, to aid in risk management. This chapter discusses our findings, and Chapter 5 discusses the framework.

Key Differences Between Mission Assurance for Rapid Acquisition and Mission Assurance for Traditional Acquisition

Using semistructured interviews, we elicited insights from government and federally funded research and development center (FFRDC) SMEs regarding MA objectives, approaches, and processes for rapid acquisition. We observed that what constitutes mission success for rapid acquisition differs from that for traditional acquisition, which is performance-focused. Further, the risk postures associated with traditional acquisition and rapid acquisition are very different (i.e., rapid acquisition is risk-tolerant and traditional acquisition is risk-averse).[36] Because of these fundamental differences, we found that the traditional framework and approach to MA are inadequate for rapid acquisition. Table 4.1 summarizes key differences in MA approaches between traditional and rapid space acquisition. We discuss the differences and implications for how MA is managed for rapid acquisition in detail in the remainder of this section.

---

[36] By *risk posture*, we are referring to the level of risk an organization is willing to accept or tolerate.

Table 4.1. Key Differences in Mission Assurance Approach Between Traditional and Rapid Acquisition Programs

MA for Traditional Space Acquisition	MA for Rapid Space Acquisition
• Focuses on system	• Focuses on warfighter/mission
• Addresses technical risks to the narrow system	• Addresses technical, operational, and programmatic risks of the broader mission
• Averse to technical risk	• Tolerant of technical risk
• Maximizes performance-centric MA objectives (mission capability and reliability) that drive cost and schedule	• Balances multiple MA objectives (schedule, mission capability, reliability, security, resilience) within cost constraints

There Are Additional Mission Assurance Objectives for Rapid Acquisition, and Their Relative Priorities Differ from Traditional Acquisition

Traditional space systems typically have been associated with long mission lifetimes with limited ability to repair once on orbit, high costs, high complexity, and critical national security missions, which drove the risk tolerance for those programs to be very low. Additionally, some of these characteristics drove traditional space programs to emphasize maximizing mission capability and stringent performance requirements.[37] As a result, MA requirements for traditional space acquisition are rigorous to minimize the probability of failure in the acquired system, subject to launch and natural space environments.[38] Such military space systems are often categorized as Class A systems, per the four risk tolerance classes (A, B, C, and D) used by the space community to convey risk profiles and their associated MA standards. Class A refers to very low risk tolerance, with the most comprehensive and rigorous MA standards ("gold standard"), and Class D reflects high risk tolerance, with minimum MA standards.[39] Appendix D provides additional details on the characteristics associated with the risk profiles for Class A to D and high-level summaries of traditional MA processes for each class.

The traditional MA standards for Class A systems are focused on technical and engineering aspects of the acquired system to assure with high confidence that the system will meet high performance requirements, to maximize mission capability, and high reliability requirements, to ensure that the system will be functional for the duration of the mission lifetime. Schedule delay and cost growth were, to some extent, tolerated in past space acquisition programs to achieve stringent mission capability and reliability goals.

However, these traditional MA objectives are inadequate for assuring mission success for rapid acquisition. There are additional objectives that are critical to ensuring mission success.

---

[37] Ellen Pawlikowski, Doug Loverro, and Tom Cristler, "Disruptive Challenges, New Opportunities, and New Strategies," *Strategic Studies Quarterly*, Spring 2012.

[38] See Johnson-Roth, 2011.

[39] National Aeronautics and Space Administration, Office of Safety and Mission Assurance, *Risk Classification for NASA Payloads*, NPR 8705.4A, April 29, 2021.

The primary reason that the USSF is pursuing rapid acquisition is to outpace threats. Assuring mission success in a space warfighting domain means that the relevant capability has to be available ahead of threats, and it has to work in and through contested space environments. PMs managing rapid programs whom we interviewed emphasized that schedule is paramount for mission success in rapid acquisition, and thus they are willing to accept more risks in mission capability or reliability to achieve the schedule goal. Many rapid programs are also budget-constrained. Thus, the MA objectives for rapid acquisition need to be expanded to reflect such operational and programmatic goals (in addition to technical goals) that constitute mission success for rapid acquisition. The objectives should include timely delivery of capability on schedule, resilience, and security to ensure that the acquisition is focused on threats as well as on the traditional MA objectives. We define these objectives as follows:

- **Schedule:** delivery of capability on an operationally relevant timeline
- **Resilience:** ability to perform functions necessary for mission success with shorter periods of reduced capability in contested space environments[40]
- **Security:** ability to reduce the likelihood of malicious or unauthorized actions that could compromise or damage critical assets necessary for mission success
- **Reliability:** ability to perform functions as expected for the duration of mission lifetime
- **Mission capability:** ability to execute missions effectively (typically associated with system technical performance).

There are inherent tensions among these MA objectives, and thus rapid acquisition requires trade-offs and striking the right balance among them. This new trade space is changing how decisions are made in rapid acquisition. For instance, our interviewees indicated that rapid acquisition is willing to accept more risks in achieving traditional MA objectives (i.e., meeting the full set of performance requirements, high reliability), while buying down the MA risk associated with fielding capability late (and not getting ahead of the threats). The implications are that traditional Class A MA should not be expected from rapid programs; performance is tradeable in rapid acquisition; and sensible failure (i.e., taking calculated risks and failing forward) should, in principle, be tolerated. This MA philosophy is fundamentally different from what is expected from traditional space acquisition that adheres to traditional MA processes.

The MA objectives in the five dimensions are driven by acceptable risk levels in each dimension, which will vary depending on program constraints (e.g., fixed schedule and budget) or operational priorities. For instance, a program may establish that lower mission capability (e.g., 80 percent of full capability) is acceptable for mission success, but delay in fielding is unacceptable because that may render capability obsolete (perhaps because the threat is evolving rapidly). The Operationally Responsive Space-1 (ORS-1) program was one such example.

---

[40] This definition is an adaptation of the definition provided in Office of the Assistant Secretary of Defense for Homeland Defense and Global Security, *Space Domain Mission Assurance: A Resilience Taxonomy*, September 2015.

Despite high technical risks it faced at launch, the risks were accepted, and the satellite was launched on schedule because the warfighter was facing an urgent need.[41]

A program office may wish to identify more granular or specific objectives to guide the trade-offs among the MA elements for the specific system and situation. As an example, an MA objective associated with mission capability could be stated as "The system needs to provide imagery over a particular area at least once a day" for the operator to achieve the mission. Or an MA objective associated with reliability might be stated as "The system should function at a minimum of 80 percent capacity 90 percent of the time, for a minimum of two years." If other objectives are critical to assure mission success, they should also be included in this trade space. The identification of these objectives is an important step toward achieving MA.

Mission Assurance for Rapid Acquisition Is Tailored Based on Risk Assessments and Trade-Offs to Balance the Objectives

We elicited information about the MA approaches used by USSF rapid programs from program office personnel and found that the MA approach for rapid programs is generally tailored and that it varies from program to program. What constitutes MA and how it will be accomplished for each rapid program are different, unlike traditional space acquisition, which has generally adhered to common, rigorous MA standards associated with Class A missions.

Additionally, several rapid program office personnel highlighted that the resources allocated to MA for rapid programs are less than those for traditional programs. Thus, each program tailors the MA requirements and process based on its risk posture, priorities, and resources.[42] However, certain rapid programs may apply the traditional MA standards for Class A missions because of their high criticality to national security, as in the case of the Next-Gen OPIR program.

The MA process can vary in terms of the rigor (e.g., breadth and depth of MA activities), the level of government involvement, and the system element to which MA is applied (e.g., the level of system assembly at which MA is applied, or the payload to which different MA standards are applied), depending on the acceptable risk levels. For example, one interviewee noted that MA activities related to parts is minimal in some rapid programs, and that it is only one area of risk, especially when working with new vendors or using commercial parts or practices.[43]

---

[41] Barbara Braun, Lisa A. Berenberg, Sabrina L. Herrin, Riaz S. Musani, and Douglas A. Harris, *A Class Agnostic Mission Assurance Approach*, Aerospace Corporation, TOR-2021-00133, January 15, 2021.

[42] Many of the rapid programs in the USSF are still in the design phase and have not reached the build or test phase. It is possible that the MA approach could change as the program progresses. Rapid programs are relatively new to the USSF, and this is a first iteration of implementing a tailored MA approach rather than following the standard Class A MA process.

[43] The MA process for parts, materials, and process include a set of activities to assure that the parts, materials, and process used in manufacturing will function as intended. These include parts and material testing for space and launch environments, review of parts tests, and so forth. See Johnson-Roth, 2011, for more details.

The PTS program, for example, does not prescribe MA requirements to the contractors but rather, that the contractors tailor the MA approach and make the MA trade-offs (e.g., a contractor may focus more on enhancing the mission capability and use low-risk components that might require less MA effort).[44] The government mainly maintains insight and provides input as necessary, as is the case with the testing. The contractor is responsible for the test plan, but the program office might provide feedback based on the test and evaluation community's recommendations. Ultimately, the contractor decides whether to implement such recommendations. The program office noted that this approach works because the contractors are incentivized to win the contract for the next phase. However, some MA practitioners cautioned that limited government oversight could lead to potential risks if the government is using contractors with less experience.

Another key difference in the MA process for rapid programs is that it is likely to be more fluid and iterative. An interviewee reflected that the sequential nature of the traditional MA approach (similar to the traditional acquisition process) can have adverse impacts on schedule. For instance, the contractor waits until gates or milestones to share information about any issues encountered during system development. Further, such issues or anomalies have to be resolved (e.g., additional testing, independent reviews, risk assessments) before proceeding forward, introducing schedule delays. Rapid programs have more flexibility in how those gates or milestones are implemented, and interviewees highlighted that the interactions with the contractors are more frequent. Interviewees highlighted that the close communication with the contractors enables them to bring up the issues as they are uncovered and resolve them early on.

This iterative characteristic is also found in agile software development. Some interviewees argued that this approach is, in and of itself, an MA approach. That is, as the program designs, develops, builds, and tests incrementally and rapidly, it discovers and resolves issues early on and hence improves MA. The same logic applies to many rapid programs that are prototypes. Lessons from prototypes will feed into improving MA for the follow-on, next-generation system that is providing the enduring capability.

While a mature MA framework and set of guidelines exist for traditional space acquisition (as a result of decades of lessons-learned and building a large body of knowledge), a standard, overarching framework for MA for rapid acquisition appears to be lacking.[45] This is not

---

[44] As Tables D.2 through D.8 show in Appendix D, there are many MA processes. For each of the MA processes, the government might levy requirements on how rigorous the MA process should be in terms of application level, technical depth and breadth, and oversight. For example, one MA requirement associated with the integration, test, and evaluation MA process might be to conduct tests at all levels of assembly (down to the component level) to ensure that functional performance and interface requirements are met.

[45] Theoretically, the traditional MA framework could be applied to rapid acquisition by applying the standards associated with more risk-tolerant mission classes (i.e., Class B, C, or D risk classes). However, there are two main shortfalls associated with that approach (see Braun et al., 2021). First, it is unlikely that the risk profile for a rapid program can be characterized by a single acceptable risk level for all aspects of the program and for all system

surprising given that inclusion of rapid acquisition (e.g., MTA) in the AAF is relatively new.[46] However, the Aerospace Corporation has developed the "class-agnostic MA" framework, which is being applied to certain Space RCO programs, SSC prototype programs, and programs in other organizations, such as NASA.[47] This framework does not provide specific guidance on MA standards and practices, unlike the traditional MA framework: Rather, it provides guidance on developing and executing a MA approach that is based on risk assessments and risk prioritization.[48] As one practitioner of the class-agnostic MA approach puts it, MA is "anything you do to improve the probability of success of a mission," indicating a vastly different philosophy on MA from that of the traditional MA framework.

A common theme among the rapid programs is that risk assessment and risk management are fundamental to assuring mission success, and that trade-offs are necessary. We observed that rapid programs are making trade-offs between schedule and other elements of MA (mission capability, security, reliability, resilience) to achieve the schedule goals. For instance, the PTS program is accepting more supply chain security risk by using field programmable gate arrays instead of trusted foundry for schedule benefits. The program is also willing to accept risks in mission capability (i.e., accept reduced performance) to achieve the schedule goal. Many interviewees supporting rapid programs expressed that they are willing to accept risk in mission capability because a partial capability delivered on time (to get ahead of the threat) is better than full capability delivered late (and hence obsolete because the threat has changed). A Space RCO mission lifetime associated with the systems being acquired may be short, and interviewees indicated that some of the programs are tolerating more reliability risk because the operators may still get some value even if the system fails (e.g., a system on-orbit could serve as a messaging mechanism and provide deterrent value). Another example of a trade-off that Space RCO makes is that it focuses on a subset of threats (presumably the highest-priority threats) rather than a

---

elements. Realistically speaking, a risk profile for a rapid program is reflected by varying degrees of acceptable risk levels associated with different elements of a system or different aspects of the program. For instance, the risk tolerance associated with the primary payload may be low while that associated with the communications subsystem might be high. Or a program might have a higher risk tolerance for reliability than that for system performance because of a short mission lifetime. Thus, a rapid program may not perfectly fit into one of the risk mission classes. Second, the traditional MA framework provides guidance on practices and standards for assuring mission success from a technical or engineering perspective. Those guidelines do not aid making trade-offs and prioritizing risks to achieve multiple MA goals for rapid acquisition.

[46] DoD has limited experience in rapidly developing and fielding space systems that are not experimental or demonstration systems. In 2007, in response to the need for rapidly fielding space capabilities to meet warfighters' urgent needs, Congress established the Operationally Responsive Space (ORS) Office, which developed and fielded four experimental or operational prototypes before Congress redesignated it as the Space RCO in 2018 (GAO, *DoD Faces Challenges and Opportunities with Acquiring Space Systems in a Changing Environment*, GAO-21-520T, May 2021b).

[47] Aerospace Corporation staff, discussion on MA for rapid programs, 2021.

[48] Refer to Braun et al., 2021, for more details.

comprehensive set to achieve the schedule goal while delivering a capability that will be sufficiently robust against a range of threats.

An Experienced Workforce and Foundational Infrastructure Are Necessary to Achieve Mission Assurance for Rapid Acquisition

Risk assessments are fundamental to rapid acquisition's MA approach. Program office personnel indicated that tailoring of the MA approach and making trade-offs between the technical dimensions of MA (e.g., mission capability, reliability) and nontechnical dimensions (e.g., schedule) are largely based on risk assessments that are informed by expert judgment.

Rapid programs rely on SMEs with deep expertise across a wide range of disciplines, and interviewees raised several related concerns they face. First, the collective knowledge in software development and cybersecurity in the acquisition community is inadequate. Although this deficiency is not unique to rapid programs, it presents a greater risk for programs that are aiming to go faster. For example, one program office staff member highlighted that the program gets "slowed down by the amount of education that people need to convert to agile [from waterfall]."

Second, there is a concern about the vulnerability of the intellectual capital. Deep expertise is built from lessons learned from past programs and from experiences in traditional programs that apply rigorous MA standards. Rapid programs thus rely on traditional programs to access that deep knowledge base. However, as more programs aim to go faster, there is a concern that the intellectual capital may atrophy—but the counter to this is that successfully managing rapid acquisition programs creates a new set of skills.

Broader Mission Assurance Risks for Rapid Acquisition

Through our interviews, we found that some stakeholder communities are lagging in adopting streamlining methods and have different risk posture and expectations for MA. Such alignment issues between the program office and the other stakeholder communities could introduce risks to assuring mission success. We also found that rapid acquisition may exacerbate mission engineering (ME)/system-of-system (SoS) integration challenges, which also imposes MA risks. We discuss these two broader MA risks in detail in this section.

Alignment Issues Among Stakeholder Communities

Achieving MA goals requires contributions from multiple organizations, not just the rapid program office or the acquisition community alone. Figure 4.1 illustrates these stakeholder communities. The user community (combatant commands and organizations representing the user community) provides threat-informed capability requirements to the acquisition community. The functional communities that are part of the broader acquisition community (e.g., contracting, systems engineering, test and evaluation, acquisition intelligence, safety) provide support for the program office. The operational community (operators who perform the mission and the organizations that represent the operators) provides inputs to the mission requirements as well, such as assuring that the acquired capability is ready for operational use (i.e., the capability "meet[s] operational requirements and have the necessary elements required to support mission execution").[49] The contractors who design, develop, and build the system are also an integral part of this ecosystem.

Figure 4.1. Alignment Challenges Among Stakeholder Communities for Assuring Mission Success for Rapid Acquisition

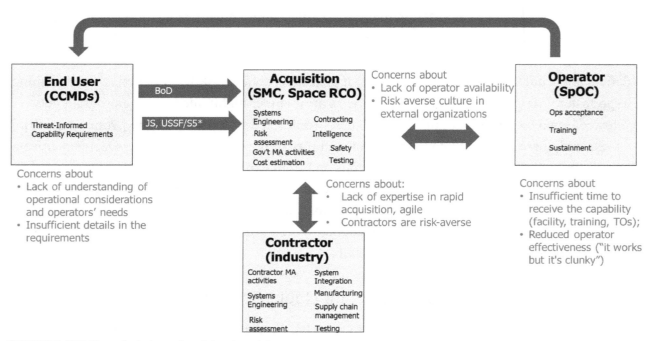

SOURCE: RAND analysis based on interview data.
NOTES: CCMDs = combatant commands; BoD = Board of Directors; JS = Joint Staff. The requirements for Space RCO programs are approved by the Board of Directors. Current USSF rapid programs initiated prior to the stand-up of the USSF and thus the requirements came from AFSPC/A5. Subsequently, Space Operations Command Headquarters picked up much of the requirements responsibilities and were working with the acquisition community on operational requirements during the course of our interviews. The requirements responsibilities transferred to the USSF Headquarters with the latest reorganization of the Space Operations Command.

However, although the broader community in the USSF recognizes the need to go faster to outpace threats and has bought into the idea of rapid acquisition, our interviews revealed that some stakeholders are lagging in adopting streamlining methods and a risk-tolerant mindset. Different expectations for assuring mission success among the stakeholder communities are creating challenges for rapid program offices to coordinate and collaborate with them to deliver a capability and to achieve the MA objectives. Figure 4.1 highlights a sample of key concerns

---

[49] Air Force Space Command Instruction 10-605, *Operational Acceptance Process*, 2016.

many interviewees raised regarding rapid acquisition that reflect alignment challenges that a rapid program is facing. We expand on these sample concerns below:[50]

- Some interviewees raised concerns that high level requirements and a short requirements document[51] associated with rapid acquisition could lead to potential problems after the program gets started because the acquisition community may not fully understand the requirements and would need to come back to the user or operator community to get more clarification.[52] Rather than clearly defining the requirements prior to the program start, requirement clarification would occur after the program starts, leading to potential inefficiencies (e.g., requirements creep, cost or schedule increase). Some interviewees believed that the acquisition community does not have an adequate understanding of operators' needs.
- The acquisition community expressed the lack of operator availability to get their operational perspectives and input on priorities. We found that the operator involvement may vary depending on the program. The Space C2 program at SSC, for example, appears to have active participation from the operator community, given the DevSecOps environment in which software is developed.
- Some interviewees raised concerns about the lack of experience in rapid acquisition (especially in agile software development) in the traditional space industrial base. They observed that some of the traditional defense contractors are lagging in risk tolerance.
- The acquisition community raised concerns that the operator community has not yet tailored their processes and has been more risk-averse to accept capabilities faster. Program office personnel highlighted challenges with getting other external organizations, such as the National Security Agency, to tailor their processes to better accommodate rapid acquisition.
- The operators have a different perspective on MA or at least different priorities. For the mission to be assured, the operators expect the system to be effectively operable and the operators' needs to be met. However, operators have raised concerns that the acquisition community is not providing all the necessary "ilities" for them to be effective. They also highlighted that the capability is being delivered too fast, and, as a result, they have inadequate time to get the necessary elements in place to be operationally ready. These elements include trained personnel, accredited facilities, funding, technical orders for

---

[50] Some of these challenges are also covered in Chapter 3.

[51] By *short requirements document*, we're referring to a requirements document that is short in length. Even if the key performance parameters (KPPs) are few in number, a lengthy requirements document could include other details that impose requirements that support the KPPs. For example, in addition to the KPPs, a requirements document may list specific details on how the KPPs are decomposed or applied to different functions of a system. It may contain details on various threat levels or operating conditions under which the KPPs should be met. Such details may constrain the PM's ability to do trades, but they may also provide important information that will help the acquirers better understand the requirements.

[52] We make the distinction here that the user and the operator communities are different, but in some cases, they could be the same community depending on the space capability area. For instance, for space capabilities in the force enhancement area (e.g., SATCOM; position, navigation, and timing [PNT]; weather), the user community receives the space effects provided by the SPACECOM. For other capability areas, such as in space control, we treat the user community as the same as the operator community.

maintenance, and others. Operators are concerned that they may end up accepting the capability at risk if these elements are not adequately in place.

These concerns point to lack of alignment in risk postures and varying degrees of readiness to support rapid acquisition across the stakeholder communities, which could introduce schedule delays or other risks to achieving MA. The challenge is that the AAF and prior emphasis on tailoring are relatively new, and many organizations are still adapting to rapid processes and mindset.

Another related challenge is that expectations of MA standards for rapid acquisition may differ among various stakeholders. Each rapid program has unique MA objectives, a unique approach to MA, and acceptable risk levels that are tailored, which could lead to potential challenges in communicating MA standards and expectations to the stakeholders external to the program office. Some stakeholders may still be expecting to meet the same degree of standards as for traditional acquisition. Some stakeholders may have a different definition of mission success (e.g., operational community versus acquisition community) and relative priorities of MA objectives. Risk posture can also change during the program life cycle for a variety of reasons—for example, because of the change in the leadership, or because leadership becomes more risk-averse as the program gets close to launch.

The relative newness of rapid acquisition and the tailoring aspect of rapid acquisition appear to make alignments of standards, processes, and practices across multiple stakeholder communities more challenging. Interviewees emphasized that key enablers to alignment are early and frequent communication with stakeholders and adequate documentation on the risk posture.

Mission Engineering/System-of-Systems Integration for Rapid Acquisition Mission Assurance

Assuring that a capability is delivered to the warfighter in a timely manner requires multiple interdependent systems to be synchronized and integrated. For instance, even if a rapid program delivers a satellite on time, the mission is not assured if the necessary ground systems are not also available to operate the satellite and process the satellite data. Although SSC has experienced ME/SoS integration challenges with traditional acquisition in the past,[53] the rapid acquisition environment further exacerbates the integration and synchronization challenges as we discuss below.

To illustrate the potential complexity in ME/SoS integration, the graphic in Figure 4.2 depicts the various interdependent programs involved in assuring delivery of missile warning capabilities to the end user and the associated program characteristics. The Next-Gen OPIR space vehicles will be delivered as a result of three separate programs, one of which is a resiliency payload delivered by an external organization. As noted earlier, rapid acquisition tends to lead to more segmented programs because it focuses on delivering a minimum viable product,

---

[53] Shelton et al., 2021.

which is not necessarily an enduring capability or the entire capability. For instance, Next-Gen OPIR is an MTA project that is delivering a primary payload, which needs to be integrated with a bus that is not part of the MTA.[54]

Figure 4.2. Mission Engineering/System of Systems for Delivering Missile Warning Capability

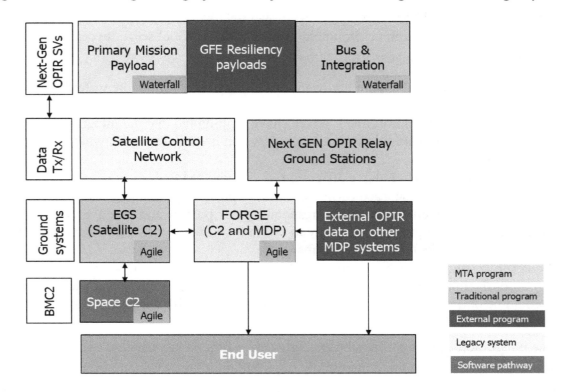

NOTES: The primary mission of the Next-Gen OPIR system is missile warning, and it has secondary missions to support missile defense, technical intelligence and battlespace awareness. SV = space vehicle; BMC2 = battle management command and control; MDP = mission data processing; Tx/Rx = transmit/receive.

The Next-Gen OPIR space vehicles also require multiple supporting ground elements for the operators to provide the critical missile warning capability. It is not unique to rapid acquisition for a satellite program to be separate from its associated ground programs. The ground programs might provide an enterprise solution or a solution that is dedicated to a particular satellite program. The Satellite Control Network and the Enterprise Ground Services support the enterprise for data transmit/receive (Tx/Rx) capabilities and satellite operations.[55] In contrast, the Relay Ground Stations and the FORGE program are dedicated to the Next-Gen OPIR satellites

---

[54] PTS is another MTA program that is delivering the primary payload, which needs to be integrated with a bus that is not part of the MTA.

[55] Satellite telemetry, tracking, and commanding (TT&C) and mission data downlink are achieved via data Tx/Rx network and dedicated ground stations.

for mission data downlink and mission data processing, respectively.[56] Another program that cuts across the enterprise is the Space C2 program, which will provide space domain awareness and battle management and C2 capabilities for all USSF satellites.

While MA in space acquisition has traditionally been a program-level function, focusing on a particular system, we assert that to truly assure mission success, MA should be viewed from an ME/SoS perspective. For this reason, we observe that the integration and synchronization challenges in a rapid acquisition environment pose MA risks. We observe several factors that are presenting integration challenges:

- Satellite programs (which are hardware-driven) typically follow the waterfall approach, whereas software-dominant programs may follow an agile approach (delivering capabilities incrementally). Interviewees have highlighted that the integration of waterfall and agile has been particularly challenging, which is consistent with what we found in our literature review.
- Each program is likely to have different risk postures, MA objectives, and MA approaches. As we discussed earlier, each rapid program is uniquely tailored. Aligning different MA objectives and the corresponding risk posture and MA approach across all the interdependent programs is likely to be complex. Such challenges could potentially introduce MA risks between the seams.
- Enterprise solutions such as Enterprise Ground Services are designed to meet the needs of all or the majority of the satellite programs. Hence it is unlikely that such programs would be optimally aligned in schedule or risk posture with any one particular satellite program.

These integration challenges may be exacerbated by limited government expertise in systems integration, limited ability to integrate rapidly (e.g., Modular Open Systems Approach is not yet established), and perhaps lack of a single organization responsible for SoS integration (or a lead system integrator) within SSC for integrating those systems that it is acquiring. Delivering space capabilities may also require coordinating interdependencies with external organizations, which carries many challenges because each organization may have different priorities, requirements, timelines, or other constraints that could lead to conflicting interests. The PIC is serving this role by coordinating such dependencies with its "mission partners," such as the National Reconnaissance Office, Space RCO, Space Operations Command, and the Space Development Agency.

Summary

In this chapter, we discussed how risks created by techniques used to streamline acquisition might impact MA. A major challenge in trying to assert or examine a causal relationship between streamlining techniques and degradations in MA is that numerous factors unrelated to

---

[56] FORGE will eventually be used to support other OPIR satellites and integrate with related ground systems from other organizations.

streamlining can affect MA. Thus, it is extremely difficult to establish causality, so we focused on how MA can be affected, broadly and even indirectly, by efforts to speed up acquisition.

From our interviews with SMEs, we learned that traditional approaches to MA are inadequate for rapid acquisition. These approaches are predicated on space programs with long mission lives; limited or no capability for on-orbit repair; high complexity, technical requirements, and costs; and criticality to national security, driving low risk tolerance. Thus, traditional MA is focused on performance aspects. The MA objectives for rapid acquisition need to be expanded to reflect operational and programmatic goals (in addition to technical goals) that constitute mission success for rapid acquisition and delivery of capability to the warfighter.

In short, the MA trade space for rapid acquisition must consider schedule, security, resilience, reliance, and mission capability, and these attributes must be balanced and driven by what is determined to be acceptable risk for each dimension. Thus, MA approaches for rapid programs must be tailored, program-specific, and iterative, rather than fixed.

Several major challenges to such an approach were identified that could heighten programmatic risks. These include a shortage of technical personnel qualified to assess risk appropriately and a misalignment in adoption of streamlining approaches, in risk postures, and in MA standards across stakeholder communities.

Space acquisition MA has traditionally been a program-level function. However, the increasing complexity of programs and systems is demanding an ME/SoS approach. The need for rapid acquisition may exacerbate cross-system integration and synchronization challenges and risks to MA.

5. A Framework for Managing Mission Assurance Risks for Rapid Programs

As discussed in Chapter 4, MA for rapid acquisition is fundamentally grounded in sound risk assessments. Achieving MA for rapid acquisition will require making trade-offs among multiple MA objectives based on risk assessments throughout the program life cycle. To that end, we propose a MA risk management framework to help balance MA objectives and to facilitate communications of risks with stakeholders.

There are five main steps in the proposed MA risk management process as depicted in Figure 5.1. These steps are generally similar to the risk management process currently being used, except that trade-offs are required among multiple MA objectives.[57] Our framework also includes the following attributes:

- a structured way of thinking about MA from day one
- a disciplined approach to making risk trade-offs to achieve mission success
- a mechanism to explicitly make trade-offs with inputs from stakeholders
- an approach to manage risks collectively, rather than mitigating individual risks.

This framework builds on existing processes and concepts with which the space industry and government personnel are already familiar, and thus it should facilitate adoption and implementation.

This chapter presents a step-by-step overview of the use of the proposed framework, followed by a notional example to illustrate the use of the framework.

---

[57] See DoD, 2017, p. 17; Braun et al., 2021, p. 9.

Figure 5.1. RAND's Proposed Mission Assurance Risk Management Framework

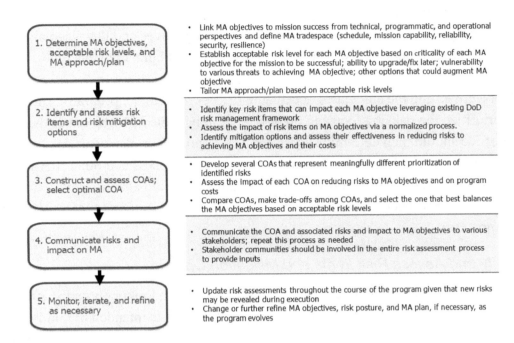

NOTE: The program office would own this process, but it is critically important that stakeholders provide inputs in steps 1 through 3. COA = course of action.

Mission Assurance Risk Management Framework for Rapid Acquisition

Step 1: Determine Mission Assurance Objectives, Risk Posture, and Approach/Plan

As a first step, the program office defines the MA objectives for the program. The MA objectives should include metrics that are linked to mission success from a technical, programmatic, and operational perspectives and capture the key elements required for the operators to perform their mission successfully. We offer the five high-level metrics discussed earlier to capture those key elements of MA and to define the MA trade space: schedule (timeliness of capability delivery), mission capability, reliability, security, and resilience. A program office can tailor these metrics or include others as appropriate to better reflect the program's MA goals.

For each MA objective, acceptable risk levels or the risk posture should be defined. Multiple factors can affect the risk posture, including

- criticality of each MA objective for the mission to be successful (individually and relative to each other to facilitate trade-offs)
- ability to upgrade or fix later
- other options that could augment MA objective
- vulnerability to various threats to achieving each MA objective.

47

The MA objectives and the acceptable risk levels should be developed in collaboration with the broader community that influences acquisition (operators, intel, engineers, etc.) to the extent practicable. The program office should communicate these objectives and the acceptable risk levels to the other stakeholders to ensure that they are realistic and that they meet the stakeholders' expectations.

The MA objectives and the associated acceptable risk levels help the program office, and the stakeholders prioritize the risks by focusing on those that impact the MA objectives the most. The program office should then determine a tailored MA approach or plan. The risk posture will drive how much rigor (breadth and depth) is needed in the MA process and other MA-related activities (e.g., systems engineering, testing). The risk posture should also help other stakeholders tailor their respective standards and processes (e.g., operations acceptance or safety standards and process).

Step 2: Identify and Assess Risk Items and Mitigation Options

The next step is to identify and assess key risks in the program, looking broadly across various risk sources that could affect the MA objectives. This step is intended to build on the risk assessment that the program office is already performing using the DoD Risk Management Framework discussed in Chapter 3.

The DoD Risk Management Framework broadly categorizes risk items in terms of technical, programmatic, and business risks. Further, these risks are assessed in terms of their impact on the traditional program objectives: performance, schedule, and cost. To apply RAND's MA risk management framework, the risks identified in the DoD framework would need to be assessed in terms of more granular risk categories that reflect the MA objectives (i.e., schedule risk, mission capability risk, reliability risk, security risk, and resilience risk). The program office may identify additional risks that were not initially identified using the DoD framework. Similar to the existing process that the program office is following, the impact of the identified risks on achieving MA objectives are subjectively assessed using inputs from SMEs (from all the relevant stakeholder communities as appropriate). A normalization process must be included in the risk assessment method to enable comparison of disparate risk dimensions on a common scale.

Once the risks are assessed via a normalization process, the program office should identify mitigation options and assess their effectiveness in reducing risks to achieving the MA goals. Additionally, the impact of each mitigation option on cost should be captured. Ideally, multiple mitigation options should be identified for each risk item. Note that it is possible that a mitigation option that decreases risk in one MA objective could increase risk in another.

Critically, Step 2 should be updated throughout the course of the program, given that new risks may be revealed during execution.

Step 3: Construct and Assess Courses of Action and Select the Optimal One

The third step of the MA risk management process is to prioritize the risks and make trade-offs to balance risks to the MA objectives since not all risks can be mitigated given schedule and resource constraints. Here, the program office develops several courses of action (COAs) that represent meaningfully different prioritization of identified risks. Each COA is a combination of risk acceptances and mitigation actions for each risk item. The program office then assesses the impact of each COA on reducing risks to MA objectives and on program costs using the risk assessment and cost information derived from Step 2. The program office would then compare the COAs to select the one that best balances the MA objectives based on the acceptable risk levels. A method for aggregating the impact of individual risks on a MA objective to arrive at the overall impact on the MA objective would need to be developed. We offer one method in our example discussed later in this chapter.

Step 4: Communicate Courses of Action and Its Associated Risks and Impact on Mission Assurance to Stakeholders

The next step is to communicate the selected COA and its associated risks and impact to MA to various stakeholders. Ideally, the stakeholder communities should be involved in the entire risk assessment process (Steps 1 through 3) to provide inputs and facilitate alignments of the risk posture and processes that affect rapid acquisition. Note that the MA objectives, risk posture, and MA plan may need to change or be further refined as the program evolves, and those changes should, again, be communicated to the stakeholders.

Documentation of risks, risk assessments, and other key factors that led to decisions about risk acceptance and COAs is also important to facilitate communications and common understanding among stakeholders.

Step 5: Monitor, Iterate, and Refine, as Necessary

As risks are uncovered, mitigated, and accepted, new risks may be revealed, or mitigations may not be as effective as anticipated. For these reasons or other unanticipated risks that may be introduced as a result of events outside the control of the program office (e.g., funding cuts), the program office will need to continually monitor risks and iterate the MA risk assessment and management process. A program office should determine the frequency of routine risk monitoring and thresholds for iterating the risk assessment and COA evaluation. The frequency and conditions for re-assessing risks could vary depending on the program office's ability to tolerate risks, the program duration, key milestones (design reviews, DT, etc.), available resources, or relative priorities of the MA objectives. Additionally, the risk assessment process may need updates when new critical information (e.g., new threat assessment) becomes available or other significant changes in the program occur (e.g., leadership change) because the acceptable risk levels may need to change as a result. Unanticipated events, such as funding cuts, discovery of anomalies, or a supply chain disruption, may also warrant repeating the risk assessment because initial assumptions about the program may no longer be valid.

An Illustration of the Mission Assurance Risk Management Framework Using the Next-Gen OPIR as an Example

In this section, we illustrate the application of our MA risk management framework by applying it to the Next-Gen OPIR program using notional risks and acceptable risk levels. We focus on the first three steps for the purpose of illustration.

Step 1: Defining Acceptable Risk Levels for Mission Assurance Objectives

In the first step of the MA risk management framework, we define the acceptable risk levels for each MA objective. The *notional* acceptable risk levels are detailed in Table 5.1. As the notional example shows, there is no specific format for describing the acceptable risk levels. The acceptable risk levels could be defined quantitatively or qualitatively. And they could be specific or generic, as long as they are sufficiently clear to communicate the risk posture and inform risk trade-offs.

Table 5.1. Notional Mission Assurance Objectives and Acceptable Risk Levels for the Next-Gen OPIR Program

MA Objective	Notional Acceptable Risk Level
Schedule	• Minimal risk tolerance • The primary payload must be delivered by 20XX, with initial operational capability by 20YY
Mission capability	• Minimal risk tolerance associated with strategic missile warning • Reduced performance affecting missile defense is acceptable for a certain set of missile threats • Reduced performance affecting other secondary missions (technical intelligence and battlespace awareness) is acceptable
Reliability	• Critical satellite safety functions should be available X percent of the time for a minimum of Y years • Critical mission capability function should be available X percent of the time for a minimum of Y years
Security	• Critical mission data should be protected for information assurance • Supply chain attack should be minimized as practical
Resilience	• Minimal risk tolerance for identified high priority threats; must meet resilience requirements associated with those threats • Resilience requirements for lower priority threats can be traded

Step 2: Identifying Risks

In the second step, we identify a *notional* set of risks associated with the Next-Gen OPIR program and other interdependent programs that collectively provide the missile warning capability. Each risk item is then assessed to determine the risk it poses to each MA objective.

Figure 5.2 details the notional risk items and the notional risk assessments. It also lists example mitigation options and their impact (notional) on reducing the risks to the MA objectives and on costs (notional).

For the purpose of illustration of this framework, we used a simple scoring and normalization method to reflect the severity of risk. The severity of risk to each MA objective is assessed on a five-point scale that ranges from 0 to 4, and we assumed that each MA objectives are of equal importance for the purpose of illustration.[58] This scoring method will facilitate aggregation of risks in the COA assessment step. Again, our simplified method is for illustration purposes only; a program office would need to develop its own method for capturing relative priorities among different MA objectives and applying a normalization process.

The risk items have different degrees of adverse impact on MA. Some of them present high risks to schedule, while others affect resilience. In our example, the impact of each risk item on the five MA objectives are assessed independently of each other. For example, a risk item that reduces reliability would be assessed in the reliability dimension only, even if such a risk could in turn introduce a mission capability risk or a resilience risk.

---

[58] Because this example is for illustrative purposes only, we did not develop rating criteria associated with the scale. Rating criteria should be established when a program office develops its own risk assessment methodology.

Figure 5.2. Example Assessment of Notional Risks in the Missile Warning Program

Risk items (Notional)	Risks to MA objectives					Mitigation cost risk
	Schedule	Mission capability	Reliability	Security	Resilience	
R1 [Subsystem]: A subsystem may not meet the design life goal	0	0	3	0	0	
Mitigation 1: Re-design/re-develop	4	0	0	0	0	2
Mitigation 2: Procure an alternative source (foreign content)	0	0	0	2	0	1
Mitigation 3: Change the subsystem design but it would increase mass and reduce performance	1	1	2	0	0	1
R2 [Mission support payloads]: Delivery of some GFEs may be delayed, potentially delaying integration	4	0	0	0	0	
Mitigation 1: Develop and execute an appropriate TTP to mitigate potential operational impact	0	0	0	0	2	0
Mitigation 2: Conduct I&T of delivered payloads as scheduled; Integrate delayed payloads late without system level testing	1	0	0	0	0	1
R3 [Performance test]: A full set of operationally relevant tests cannot be conducted to verify and validate effectiveness of mission capability	0	4	0	0	0	
Mitigation 1: Develop M&S	0	0	0	0	1	3
Mitigation 2: Delay testing until testing infrastructure is built	4	0	0	0	0	0
Mitigation 3: Develop and execute an appropriate TTP to mitigate potential operational impact	0	0	0	0	2	0
R4: [FORGE operational certification] FORGE may not be operationally certified by need date	1	0	0	0	0	
Mitigation 1: Use SBIRS legacy ground system or another agency's ground system	0	0	0	0	0	2
Mitigation 2: Get a waiver from USSTRATCOM	0	0	0	4	0	0
R5 [International agreement] Necessary international agreements may be delayed	3	0	0	0	0	
Mitigation 1: Use a stop-gap measure	0	2	0	0	0	1
Mitigation 2: Identify an avoidance measure or alternative agreements	1	0	0	1	0	0

NOTES: 0 = low risk; 1 = some risk; 2 = moderate risk; 3 = high risk; 4 = extremely high risk. GFE = government-furnished equipment; I&T = integration and testing; M&S = modeling and simulation; ITW/AA = Integrated Tactical Warning and Attack Assessment; NGG = Next-Gen OPIR GEO; ILC = initial launch capability; USSTRATCOM = U.S. Strategic Command; RGS = relay ground station; TTP = tactics, techniques, and procedures.

Taking the first notional risk item as an example, the risk a particular subsystem may not meet the design life goal is assessed to pose a high reliability risk. Three mitigation options are identified that have different impacts on the MA objectives. One mitigation option is to redesign the subsystem in question to assure that the design life goal would be met, but this would incur extremely high schedule risk and moderate cost risk. The second mitigation option is to procure an alternative source that has a more mature design. However, because of the foreign content associated with it, the security risk is assessed to be moderate. The third mitigation option would involve making minor changes to the subsystem design such that dependent components can use alternate capabilities for perhaps reduced performance, but one that is tolerable until a full solution is available. This option may reduce the reliability risk a bit; however, other risks may increase, such as schedule and in other mission performance areas.

A similar process is applied to the remaining risk items to assess the risk to the MA objectives, identify mitigation options, and assess the impact of the mitigation option on the MA objectives. This process enables the program office to explicitly keep track of how the risks and mitigation options affect each MA objective and necessitate trade-offs among the MA objectives.

Step 3: Developing and Assessing Courses of Action

The third step is the development and assessment of the COAs. In our example, we constructed three COAs, as shown in Figure 5.3. In each COA, each risk item is managed by either accepting the risk (i.e., do nothing) or by applying a mitigation action. A mitigation action is intended to reduce the risk to MA objectives to zero (low risk). However, a mitigation action that reduces the risk to one MA objective (e.g., mission capability) may increase risk to another MA objective (e.g., schedule). To determine the impact of each COA on the risk to each MA objective, we aggregate the risk "scores" associated with each risk item based on how each risk item is managed (Figure 5.4). For instance, in the case of the first COA, the aggregated schedule risk score is the average of the schedule risk score for R1, schedule risk score for mitigation 1 applied to R2, schedule risk score for mitigation 3 applied to R3, schedule risk score for R4, and schedule risk score for mitigation 1 applied to R5.[59] The relative importance of each risk item is not captured in our example aggregation method. A program office may wish to use weights for each risk item to capture relative importance (e.g., R4: FORGE might be more critical to the mission than R1: subsystem) or use an alternative method to prioritize the risk items.

The aggregated scores provide a collective look at the MA risks for each COA. Using these aggregated risk scores and the acceptable risk levels defined in Table 5.1, we select the first COA to be the best option. We note, however, that if weightings were to be applied to each MA objective, a different COA might be more desirable.

---

[59] We use averaging as a simple aggregation method for the purpose of illustrating our framework, and thereby reflecting that each risk item is of equal importance. The program office could employ a different aggregation method (e.g., applying weights to each risk item).

Figure 5.3. Example Courses of Action for Managing Notional Risks Associated with the Missile Warning Program

Risk items [Notional]	COA 1		COA 2		COA 3	
	Accepted risk	Mitigation options	Accepted risk	Mitigation options	Accepted risk	Mitigation options
R1 [Subsystem]	Yes	N/A	No	Change design	No	Alternative source
R2 [Mission support payloads]	No	TTP mitigation	No	Partial system-level test	Yes	N/A
R3 [Performance test]	No	TTP mitigation	No	Develop M&S	No	Delay testing
R4 [FORGE operational certification]	Yes	N/A	Yes	Use another agency's ground system	Yes	Get a waiver
R5 [International agreement]	No	Use stop-gap measure	No	Alternative agreement	Yes	N/A

Figure 5.4. Example Assessment of Courses of Action for Managing Notional Risks

COA	Schedule risk	Mission capability	Reliability	Security risk	Resilience risk	Cost risk
COA 1	0.2	0.4	0.6	0	0.8	0.2
COA 2	0.6	0.2	0.4	0.2	0.2	1.4
COA 3	3.2	0	0	0	0	0.4

Summary

In this chapter, we unveiled a five-step framework for assessing and managing MA risk:

1. Identify MA objectives, assess risk posture, and develop a MA approach and plan.
2. Identify and assess key risk items and mitigation options.
3. Prioritize risks, establish trade-offs, construct and assess COAs, and select optimal COAs.
4. Communicate risks and their impacts to MA to stakeholders to establish buy-in.
5. Monitor risks and repeat Steps 1 through 4 as needed.

The process outlined in Steps 1 through 5 is intended for individual programs, but a similar process may need to be established to assess and monitor risks at an enterprise level and ensure that individual program office assessments and decisions are acceptable within the broader contexts. This proposed framework should be viewed as a prototype that rapid program offices could build on and modify as they learn from its actual application Although this framework is developed with rapid acquisition in mind, it should be extensible to traditional programs.

6. Conclusions

The United States faces potential adversaries that have demonstrated increasingly effective counterspace capabilities—this was a key reason for standing up the USSF as a separate service. To outpace adversary threats in space, the USSF is pursuing rapid acquisition of warfighting capabilities.

Streamlined acquisition is not a new concept. Alternative urgent and rapid acquisition approaches have been available to meet warfighter needs for decades, and special acquisition organizations have been set up to facilitate them. Further, tailoring even traditional "waterfall" acquisition programs has always been possible (and legal under the FAR) to get capabilities at the proverbial "speed of need." The implementation of the AAF offers several pre-tailored pathways aimed at accelerating acquisition, and the USSF has taken advantage of these, particularly the MTA Pathway. The USSF has also stood up the Space RCO to facilitate the rapid delivery of new capabilities to the warfighter. The novelty of these approaches and concern about risks led the USSF to ask us to assess whether there were any risks to MA related to streamlined acquisition.[60]

Addressing this question presents several challenges, as we have described earlier. In particular, it is early in the life cycle of rapid programs to judge outcomes, and the complexities of acquisition make it difficult to link specific acquisition decisions to MA outcomes. Despite these challenges, we are able offer a set of observations about managing MA risk for rapid programs and identifying and managing potential risks to MA from streamlined acquisition. All these observations lead up to and support the important finding that MA for rapid acquisition must be considered within a trade space that includes mission capability, reliability, resilience, security, and schedule to ensure mission success.

Rapid Versus Traditional Mission Assurance Risk

The starting point in our assessment is understanding the key differences between streamlined and traditional acquisition as it relates to MA risk. The critical difference, from an MA standpoint, lies in the contrasting priorities for traditional and rapid programs. Traditional programs tend to be large and expensive, and designed to have long lifetimes. MA standards for these programs focus on technical and engineering aspects of the acquired system to assure with high confidence that the system meets high performance requirements to maximize mission

---

[60] While we were not able to fully explore the risks either as those caused by rapid acquisition or as those exacerbated by it within this research, linking the risks to different outcomes and mapping out how rapid acquisition might reduce schedule or other risks, but increase others, would be an important next step for the space acquisition community.

capability and increased reliability requirements, to ensure that the system achieves top performance during its lifetime. However, these traditional MA objectives are inadequate for assuring mission success in a contested space environment.

To outpace existing and emerging threats in an increasingly contested environment, traditional MA approaches are inadequate, and rapid acquisition principles must be adopted. While considerations and priorities of traditional MA are important, they cannot come at the expense of speed. A rapid program that delivers too slowly to meet the threat fails a key measure of mission success. Both operational and programmatic requirements must be added to technical ones when considering what meets mission success in contested environments. As described in Chapter 4, we contend that PMs must balance objectives in a broader MA trade space, which should include timely delivery of capability on schedule, resilience, and security to ensure that the acquisition is focused on threats as well as on the traditional MA objectives. This represents a new approach to evaluating MA risks.

A Framework for Balancing Risks to Mission Assurance

The need to balance among multiple objectives was corroborated by our interviews. USSF PMs are approaching the management of MA risk in a tailored way, balancing challenges based on the characteristics of the program, and taking a more fluid and iterative approach to support schedule constraints. Therefore, we propose that the USSF seek to disseminate these best practices through guidance on a formal framework that codifies existing best practices pulled from multiple programs for identifying risks to MA in rapid acquisition and determining acceptable programmatic trade-offs. We presented some key elements of such a framework in Chapter 5 (Figure 5.1). It is important to recognize that these best practices involve not only the USSF PMs but also communities beyond acquisition and the integration with other programs' capabilities to assure successful warfighting outcomes.

While USSF PMs can directly apply the concepts of the MA framework from Chapter 5, it may be beneficial for elements of the USSF (e.g., the Space RCO and/or the SSC) or the USSF as a whole to adopt, embellish, and maintain more formal guidance based on this framework. Also, given that use of the framework implies coordination with user communities and programs outside an immediate PM's control, support from the PM's organizational leadership may be necessary to facilitate these relationships. Feedback from these external user communities may also help to refine and evolve the framework to reflect practical realities and needs from a user's perspective. This may also involve and inform mission engineering and SoS engineering activities being formalized within and across DoD. Finally, metrics must be considered. The continued development and refinement of guidance on a formal framework would be greatly informed by continued collection and analysis of program data on risks, interdependencies, mitigation approaches, and effects on mission. PMs and PEOs should ensure that MARs are used to centrally report and share such information.

Addressing the Risks of Rapid Acquisition

Understanding how the risks engendered by rapid approaches to acquisition may affect MA paves the way for addressing the risks themselves. Our research identified an array of challenges relating to rapid acquisition that might affect MA, several of which we have alluded to above. (See Chapter 3 for more detail on the individual risks and mitigations). Some of these risks are common to both traditional and rapid acquisition, but introduction of shorter timelines for delivery of capabilities makes the risks more salient.

These risks (which are described in more detail in Chapter 3) include the following:

- insufficient alignment and coordination between the acquisition and operations communities, including in the adoption of processes necessary to support effective streamlining
- unreliable or inadequately timed resources
- a shortage of on-site cybersecurity experts and intelligence personnel colocated in program offices (and the overall collective knowledge of software development and cybersecurity is inadequate)
- a lag in development of needed test capabilities and infrastructure
- challenges in aligning legacy and modern software development practices, including software development life cycles
- failure to consider and plan for systems integration, particularly regarding mission engineering and system-of-systems integration
- alternate requirements processes that might specify capabilities that cannot be acquired on their rapid schedule.

Each of these risks can, itself, threaten MA, but the final and probably overarching concern is that traditional MA processes may not prioritize the accelerated schedules of programs. Identifying and addressing these sources of risk and their potential impact on MA from the outset should likely avert negative impacts on MA.

We have identified some key actions for the USSF PMs and leadership to address these risks:

- Expand the MA objectives for rapid acquisition to reflect the focus on operational and programmatic goals on top of technical goals that constitute a newly reimagined definition of mission success for rapid acquisition.
- Address the risks associated with rapid acquisition identified above. These problems are not unique to this context, and they each have identifiable mitigations. In some cases, these may be difficult to execute and require senior leadership support for change.
- Ensure that USSF processes beyond acquisition are updated to address the need to onboard capabilities more quickly. As these issues cross organizational boundaries, the acquisition community cannot address all the challenges itself.
- Proactively manage mission effects associated with rapid acquisition. We propose an MA risk assessment framework and management process (Figure 5.1) that provides a structured way to conceptualize MA from program inception; provides an approach for making intelligent risk trade-offs and choosing courses of action, with stakeholder input,

to ensure mission success; and offers an approach to manage risks collectively, rather than mitigating each one individually.

Some of these recommendations—such as alignment between different USSF communities—are addressable or even resolvable within the USSF. Some recommendations relating to specific mitigations—such as limitations of key personnel or unreliable (or inadequately timed) resources—are longstanding DoD-wide challenges that may require organizational or process redesign or even support from Congress. What they all require is senior leadership recognition of—and attention to—the issues, a focus on early and frequent communication across stakeholder communities, and a plan for—and sustained attention to—implementing change.

Appendix A. Snapshot of "Urgent" and "Rapid" Acquisition

DoD's traditional approach to developing and building weapon and information systems has been criticized for taking too long and costing too much. Multiple acquisition process solutions have been developed to go faster. The use of the term *rapid acquisition* to describe multiple pathways in the AAF is the latest instantiation of rapid acquisition. The terms *urgent* and *rapid* have been used in acquisition in several ways since 1994. This appendix provides a snapshot of the origins of rapid and urgent acquisition.

1994–2001

DoD's Advanced Concept Technology Demonstration (ACTD) Program started in approximately 1994. Military services and defense agencies adapted new but mature technologies to build prototype equipment that met a critical military need. The systems would then go to a unified command or service for evaluation in the field. An ACTD project would last two to four years. After that time, the system would enter the formal acquisition process if larger quantities were needed.[61] One such example is Global Hawk (1995), which is a high-altitude, long-endurance unmanned aerial vehicle system capable of providing broad-area surveillance. Global Hawk started as an ACTD and then became an MDAP.[62] Global Hawk has been used extensively in operations throughout the world.

2002–2014

In 2002, rapid acquisition started being used across the DoD because of the urgent warfighter needs that surged in Afghanistan and then in Iraq. Congress took a more active role, given that additional funding was needed, and requested that the Secretary of Defense create some processes for rapid acquisition:

> SecDef shall prescribe procedures for the rapid acquisition and deployment of items that are—(1) currently under development by the DoD or available from the commercial sector; and (2) urgently needed to react to an enemy threat or to respond to significant and urgent safety situations.[63]

---

[61] Congressional Budget Office, "CBO Memorandum: The Department of Defense's Advanced Concept Technology Demonstrations," September 1998.

[62] Col. G. Scott Coale, and George Guerra, "Transitioning an ACTD to an Acquisition Program: Lessons Learned from Global Hawk," *Defense AT&L*, September–October 2006.

[63] Public Law 107-314, Bob Stump National Defense Authorization Act for Fiscal Year 2003, Section 806, December 2, 2002.

During this time period, rapid acquisition intended to provide a "75-percent solution" within two to 24 months that may sacrifice affordability, interoperability, and operational suitability in order to field an effective capability more quickly.[64] One important example of rapid acquisition during this time frame (2007) was in response to the widespread use of improvised explosive devices:[65] "Thousands of Improvised Explosive Devices were killing US soldiers in Iraq and Afghanistan who were riding around in flat-bottomed Humvees using sandbags on the floor to try to protect themselves from IED blasts. It didn't work."[66] The Mine Resistant/Ambush Protected Vehicle program (MRAP) provided a solution to this critical problem. This was the first major military acquisition to go from a decision to buy to production in less than one year since World War II.

Also, during this period, DoD continued to increase its focus and policies on tailoring the acquisition processes to enable faster acquisition (among other things).

2015–Present

Rapid acquisition again changed starting in 2015, reflecting DoD's and Congress's concerns that the timelines on most weapon systems are too long. This movement has led to continued tailoring in policy and ultimately to what is currently the Adaptive Acquisition Framework. AAF is meant to develop acquisition strategies and employ acquisition processes that match the characteristics of the capability being acquired.[67] Congress also introduced the MTA Pathway in the National Defense Authorization Act for Fiscal Year 2016:

> Middle Tier of Acquisition. Rapid prototyping pathway: field a prototype that can be demonstrated in an operational environment and provide for a residual operational capability within 5 years of the development of an approved requirement. . . . Rapid fielding pathway: begin production within 6 months and complete fielding within 5 years of the development of an approved requirement.[68]

To date, the MTA Pathway has led to rapid development areas such as in hypersonics (e.g., Air-Launched Rapid Response Weapon [ARRW] Hypersonic Missile).

DoD and Congress followed up the MTA Pathway with the Software Acquisition Pathway, which is meant to ensure that DoD is using the most up-to-date, efficient software practices:

---

[64] Jim Farmer, "Hidden Value: The Underappreciated Role of Product Support in Rapid Acquisition," *Defense AT&L*, product support issue, March–April 2012, p. 46.

[65] U.S. Government Accountability Office, *Defense Acquisitions: Rapid Acquisition of MRAP Vehicles*, GAO-10-155T, October 8, 2009.

[66] Jen Judson, "30 Years: MRAP—Rapid Acquisition Success," *DefenseNews*, October 25, 2016.

[67] Defense Acquisition University, "Adaptive Acquisition Framework Pathways," webpage, undated.

[68] Public Law 114-92, National Defense Authorization Act for Fiscal Year 2016, Section 804, November 25, 2015.

Software acquisition. Provide for the efficient and effective acquisition, development, integration, and timely delivery of secure software; demonstrate the viability and effectiveness of such capabilities for operational use not later than 1 year after the date on which funds are first obligated to acquire or develop software.[69]

The Software Acquisition Pathway is relatively new, and most of the software programs are still trying to complete the planning phase.

---

[69] Public Law 116-92, National Defense Authorization Act for Fiscal Year 2020, Section 800, December 20, 2019.

Appendix B. Additional Discussion on Streamlining Practices

In Chapter 2, we provided some summary-level acquisition streamlining practices that we collected from interviews with space SMEs. In this appendix, we provide additional accelerated acquisition strategies and tactics currently being used. Table B.1 provides a complete list.

The tactics being used share many similarities. For example, at the highest level, tailoring of the acquisition process can be found across the USSF organizations and in the literature. In addition, there is some common use of solutions less than 100 percent (i.e., 75 percent or 80 percent) with the goal of reaching 100 percent later in the acquisition life cycle. These solutions would also have a narrower set of requirements. Another similarity is the use of COTS and heritage technology to reduce development time or modify proven technology. Using proven technology may also reduce production and manufacturing time.

Within engineering, there are tactics involving rapid prototyping to save time, as well as digital engineering. Likewise, the testing community is working on bringing some testing in-house, while also conducting testing in parallel with other parts of the acquisition process. This also includes meeting with testing representatives early (including DOT&E) while planning the program to mitigate risk of future schedule delays.

The contracting piece of acquisition streamlining seems to be pretty mature. The use of OTA, FAR Part 16.5 (IDIQ), FAR Part 12 (COTS), and a consortium of companies to choose from are common methods being used.

For industrial base and supply chain management, meeting with industry regularly is one way of making sure that the USSF knows what is in the pipeline in terms of technology.

A common theme in the interviews and the literature was the training and education of the workforce that is implementing the acquisition streamlining. Space RCO and SSC are both trying to attract and training a highly skilled, agile force. This is particularly true in areas where there are deficits in the number of people with skills, such as cyber, intelligence, and software acquisition.

Integration and synchronization have historically been challenges for the space acquisition community. There was a focus on building end-to-end capability, while also looking for opportunities to standardize equipment at the enterprise level.

Transition to fielding and sustainment is another area where streamlining is needed. There are several options for streamlining in this area. One important method is to secure intellectual property rights up front, as needed, before sustainment. The lack of intellectual property rights has been a persistent problem for DoD because DoD may not have intellectual property or data rights when it needs them during sustainment.[70] In addition, currently there is a major push to bridge acquisition and operational communities by using transition cells and liaisons to interface with the user community, and to collect user community feedback earlier in the process.[71]

From an organizational perspective, there are strategic differences between Space RCO and SSC, so their structures are different. Space RCO is focused on building small teams with a narrow chain of command and highly skilled embedded support. SSC is much larger and, although it is also trying to attract highly skilled talent, it has set up better communication mechanisms across the organization. For example, SSC has risk boards that meet regularly to assess programmatic risks.

Finally, acquisition streamlining relies on leadership that supports a risk-tolerant environment to improve speed (more acceptability of risk).[72] Both Space RCO and SSC noted that they have support from leadership to be allowed to fail and are learning a lot from failing often and failing fast.

Table B.1. Interviewees Identified Accelerated Acquisition Strategies and Tactics Currently Being Used in the USSF

Acquisition Processes and Functions	Accelerated Acquisition Strategies and Tactics		
	Space Rapid Capabilities Office	Space Systems Command	Literature
Acquisition process: pre- or post-AAF	• Tailoring using traditional acquisition processes (pre-AAF) • Substantial use of urgent capabilities best practices, but not using the Urgent Capability Acquisition Pathway per se	• AAF: MTA (rapid prototyping for hardware-intensive) • Software Acquisition Pathway and continuing to use agile development for software-intensive	• Tailor-in processes and documents based on the unique needs of program, as opposed to a universal standard • Make decisions often using in-process reviews, not just at Milestone A/B/C

[70] Frank Camm, Thomas C. Whitmore, Guy Weichenberg, Sheng Tao Li, Phillip Carter, Brian Dougherty, Kevin Nalette, Angelena Bohman, and Melissa Shostak, *Data Rights Relevant to Weapon Systems in Air Force Special Operations Command*, Santa Monica, Calif.: RAND Corporation, RR-4298-AF, 2021.

[71] RAND interviews with Space RCO in March 2021 and Operators in April 2021.

[72] RAND interviews with Space RCO in May 2021 and SSC in March 2021.

Acquisition Processes and Functions	Accelerated Acquisition Strategies and Tactics		
	Space Rapid Capabilities Office	Space Systems Command	Literature
Requirements	• Define 80 percent solution using narrow set of requirements • Aim for fielding on short timeline without prototyping • Use alternative requirements process (validated by SPACECOM and assigned by the Board of Directors)	• Use narrow set of requirements • Use alternative requirements process (in context of MTA prototyping) • For software, define minimal viable product to be delivered	• 75 percent or 80 percent solution better now than 100 percent solution in five to ten plus years • Follow developmental operations (DevOps) process of incorporating real-time operational feedback to generate requirements • Consider reducing requirements for later increments to lower risk • Achieve "militarily useful increment of capability" versus "desired capability"
Resources	• Use Head of Space RCO spending authority, which reduces bureaucratic layers for resource approval	• For MTA, manage within the fewer resources available than traditional Major Defense Acquisition Program programs (i.e., cost is fixed)	• Line up funding, so do not rely on reprogramming requests or highly selective prototyping funds • Have a defensible budget
Research and development	• Use COTS and heritage technology to reduce development time • Modify proven technology	• Use COTS and heritage technology to reduce development time • Modify proven technology • Conduct development and design concurrently • Consider whether the labs have something that is ready to transfer to SSC • Monitor and use innovation in industry	• Partner and collaborate across services, PEOs, Office of the Secretary of Defense, and industry to find solutions • Use agile development to "specify a little, build a little, iterate" • During planning, use a streamlined, core set of needs, strategies, and estimates before beginning development, and then iterate on them throughout program execution, which enables rapid entrance into execution, and iterative software deliveries and value assessments

Acquisition Processes and Functions	Accelerated Acquisition Strategies and Tactics		
	Space Rapid Capabilities Office	**Space Systems Command**	**Literature**
Engineering	• Use multiple iterations with 80 percent solution	• Use multiple iterations with 80 percent solution (i.e., evolution of prototypes) • Use rapid prototyping to burn down technical risk • Apply digital engineering to rapid prototypes • Use model-based systems engineering tools to generate test requirements documents • Use higher technology readiness levels	• Have government retain responsibility for final integration of mission equipment packages • Conduct "build-measure-learn" cycles • Complete a system-level preliminary design review prior to system development (U.S. Government Accountability Office [GAO] recommendation: 47 percent less unit cost growth and 35 percent less schedule growth) • Release at least 90 percent of design drawings by critical design review (GAO: 51 percent less unit cost growth and 40 percent less schedule growth)
Test and evaluation	• Bring some testing in-house • Conduct testing in parallel with other parts of acquisition process	• Conduct testing in parallel • Use an independent third-party to determine that the code is usable • Use a smaller set of code to lower testing risk	• Align test strategies with program phases to reduce risk • Meet with testing representatives early (including DOT&E) while planning program to mitigate risk of future schedule delays • Test system-level integrated prototype by critical design review (decreases schedule by 30 percent) • Reduce testing based on information gathered from the development system and previous production systems • Maximize use of automated software testing and security accreditation • Align test and integration with the overarching system testing and delivery schedules

Acquisition Processes and Functions	Accelerated Acquisition Strategies and Tactics		
	Space Rapid Capabilities Office	Space Systems Command	Literature
Production, quality, and manufacturing	• Use COTS and heritage technology to reduce production time • Modify proven technology to reduce production time	• Use COTS and heritage technology to reduce production time • Modify proven technology to reduce manufacturing time • Use existing technology from another program to reduce production time	• Use COTS and heritage technology to reduce production time • Modify proven technology to reduce manufacturing time • Use existing technology from another program to reduce production time • Use commercial parts and processes • Dual-use of COTS components, technology, manufacturing capabilities, non-traditional suppliers, and small businesses to achieve ambitious cost objectives • Use lean processing and production flow
Contract administration	• Employ business intelligence function to help maintain connections with industry and more up-to-date information • Use OTA, Multi-award OTA (new) • Use FAR Part 16.5 (IDIQ) • FAR/Defense Federal Acquisitions Regulation Supplement (DFARS) (other various parts support speed) • Has contracting authority and delegates to contracting officers (within Space RCO)	• Competition with multiple vendors (FFP) during MTA prototyping to maintain schedule, cost, and avoid vendor lock • Use OTAs and IDIQ during MTA • Use Space Enterprise Consortium (9+ companies)	• Compete the OTA to broaden the solution set • If there is potential for a successful prototype being scaled to production, consider including the appropriate follow-on language in an Other Transaction (OT) vehicle to enable award of a follow-on production OT or FAR contract without further competition • Tying milestone payments to key deliverables as an incentive mechanism • Use FAR Part 12 (COTS) • Modify existing contracts to rapidly procure COTS components • Acquire waivers to purchase equipment using sole-source contracts for various components • Use undefinitized contract actions for urgent needs
Industrial base and supply-chain management	• Use business intelligence function to provide improved or more current knowledge of contractors in the industrial base • Meet regularly with peers across industry	• Requires that the DAF have an up-to-date knowledge of contractors, suppliers, and industrial base for rapid acquisition	• Government permits contractor to use commercial standards and practices in systems engineering, test, and management • Increase contractor design responsibility and management authority

Acquisition Processes and Functions	Accelerated Acquisition Strategies and Tactics		
	Space Rapid Capabilities Office	Space Systems Command	Literature
Training and education	• Focuses on highly skilled and agile workforce for acquisition and support functions • Leans forward on training and education for the workforce	• Need to ensure good training programs because there is large turnover in software workforce	• Bring together experts from stakeholder communities early • Use independent workers and those who think "outside of the box" • Use AF Software Engineering Centers (Kessel Run) and seek out places that can help ideas grow (SOFWERX, AFWERX, CYBERWERX)
Integration and synchronization	• Focus on building end-to-end capability	• Enterprise effort to standardize bus internally and across enterprise; PIC • Use common interfaces that industry uses to rapidly upgrade the system	• Field solutions that are compatible with allied nation solutions; technology should not interfere with allied nation solutions • Employ dual contractor team under a "Lead Systems Integration" contract for "collaborative" industry/government overall management of program—key oversight responsibilities fell to contractor team
Transition to fielding/sustainment	• Transition cell in place to bridge acquisition and operational communities • Liaisons available to interface with the user community	• Secure Intellectual Property Rights up front as needed	• Have sustainment infrastructure in place • Have a maintenance plan once fielded for urgent needs • Government owns technology baseline for rights to its broad infrastructure/framework; industry for smaller efforts • Use program that DoD already owns rights and data for urgent needs • Work with user community to provide a lot of feedback • Incrementally deliver to the user • Establish User Agreements to establish governance processes that will provide feedback on minimum viable products

Acquisition Processes and Functions	Accelerated Acquisition Strategies and Tactics		
	Space Rapid Capabilities Office	Space Systems Command	Literature
Organizational structure and culture	• Use a lean structure with short/narrow chain of command • Small teams for each program • Small workforce • Embedded functional support • Constant communication with leadership • Schedule is the highest priority to counter adversary's capability	• Have risk boards that meet regularly to assess programmatic risks	• Need to use organizations within DoD where ideas are permitted to grow
Leadership support	• Leadership supports risk-tolerant environment to improve speed (more acceptability of risk) • Leadership supports figuring out innovative ways to move forward	• Has support from leadership to be allowed to fail	• Have advocacy/support for approach from senior leadership (e.g., Secretary of Defense, congressional support [key]) • High priority solution • Effectively communicate priority of urgent need

SOURCES: RAND discussions with SMEs; Defense Acquisition University Powerful Examples Library, "Powerful Example: B-52 Commercial Engine Replacement Program—Breaking Down Silos to 'Go Faster with Rigor,'" webpage, July 23, 2019; Defense Acquisition University, Powerful Examples Library, "Powerful Example: USSOCOM Supporting the Hyper Enabled Operator with Agile Logistics Capabilities," webpage, May 6, 2019; Defense Acquisition University, Powerful Examples Team, "Powerful Example: JRAC Helps Warfighters Overcome Urgent Threat from Enemy Drones," webpage, August 6, 2019; Defense Acquisition University, "Middle Tier of Acquisition (MTA): MTA Tips," webpage, undated; Defense Visual Information Distribution Service, "Adaptive Acquisition Framework and Software Pathway," webpage, January 29, 2020; Douglas Burbey, Mindy Gabbert, and Kathryn Bailey, "A Happy Medium: Middle-Tier Acquisition Authority Features Flexible Prototype and Fielding Options," *Army ALT Magazine*, September 5, 2019; Lauren A. Mayer, Mark V. Arena, Frank Camm, Jonathan P. Wong, Gabriel Lesnick, Sarah Soliman, Edward Fernandez, Phillip Carter, and Gordon T. Lee, *Prototyping Using Other Transactions: Case Studies for the Acquisition Community*, Santa Monica, Calif.: RAND Corporation, RR-4417-AF, 2020; Project Smart, "The Standish Group Reports: CHAOS," reprinted in 2014 with permission from The Standish Group, 1995; Richard H. Van Atta, R. Royce Kneece, Jr., and Michael J. Lippitz, *Assessment of Accelerated Acquisition of Defense Programs*, Institute for Defense Analyses, P-8161, 2016; Sean Brady, Office of the Deputy Assistant Secretary of Defense for Acquisition Enablers, "Faster Is Possible: DoD Publishes New Software Acquisition Policy," Defense Acquisition University, October 8, 2020; Shara Williams, Jeffrey A. Drezner, Megan McKernan, Douglas Shontz, and Jerry M. Sollinger, *Rapid Acquisition of Army Command and Control Systems*, Santa Monica, Calif.: RAND Corporation, RR-274-A, 2014; Tony Romano and Jim Whitehead, "Powerful Example: Army IVAS [Integrated Visual Augmentation System] Brings Together the Right Requirements With the Right Acquisition Strategy," Defense Acquisition University, webpage, January 27, 2020; GAO, 2020; Yool Kim, Elliot Axelband, Abby Doll, Mel Eisman, Myron Hura, Edward G. Keating, Martin C. Libicki, Bradley Martin, Michael E. McMahon, Jerry M. Sollinger, Erin York, Mark V. Arena, Irv Blickstein, and William Shelton, *Acquisition of Space Systems, Volume 7: Past Problems and Future Challenges*, Santa Monica, Calif.: RAND Corporation, MG-1171/7-OSD, 2015.

Appendix C. Additional Discussion on Risks

We gathered a lot of information on risks related to rapid acquisition from multiple sources. In order to draw out themes, we needed a framework around how the interviewees and literature discuss risk. As mentioned in Chapter 3, we used the DoD Risk Management Framework provided in DoD's *Risk, Issue, and Opportunity Management Guide for Defense Acquisition Programs*, so that the interviewees have a common framework to discuss risks. This framework is used by the PMs, PEOs, and the Secretary of the Air Force for Acquisition in the Monthly Acquisition Reports collected within the Program Management and Retrieval Tool information system. In other words, this framework exists, and the DoD workforce is familiar with it, so we are using the same terminology in this analysis. Figure C.1 provides the DoD Risk Management Framework:

Figure C.1. DoD Risk Management Framework

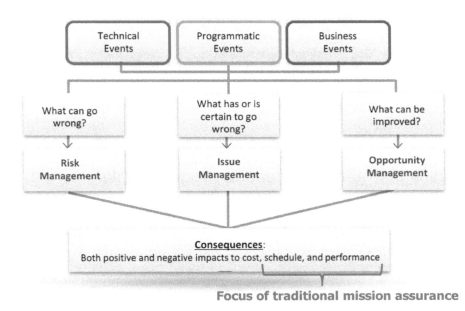

SOURCE: DoD, 2017, p. 3.
NOTE: Red, green, and purple outline is added to help the audience differentiate the three categories of events.

The risk management framework explains that there are technical, programmatic, and business events that may lead to risks, issues, or opportunities, each with cost, schedule, or performance consequences. We use these broad categories in the tables in this appendix to show that some events are more likely to impact the operational mission, and therefore, MA. These are technical events. Other events are more likely to affect the acquisition program (e.g., cause a

69

schedule delay) but not the operational mission. These are programmatic events. We also bucket the various acquisition functional areas within these three major event areas. For example, if interviewees talked about requirements creep, this would fall under "technical," while discussion of contract structure would fall under "programmatic." Finally, any discussion of user community, suppliers, etc., falls under "business." See Table C.1 for additional information on what is in each of the three categories.

Table C.1. Technical, Programmatic, and Business Events Defined

Technical	• Risks that may prevent the end item from performing as intended or from meeting performance expectations • Can be internally or externally generated • Typically emanate from areas such as requirements, technology, engineering, integration, test, manufacturing, quality, logistics, system security/cybersecurity, and training
Programmatic	• Nontechnical risks that are generally within the control or influence of the PM or Program Executive Office (PEO) • Can be associated with program estimating (including cost estimates, schedule estimates, staffing estimates, facility estimates, etc.), program planning, program execution, communications, and contract structure
Business	• Nontechnical risks that generally originate outside the program office or are not within the control or influence of the PM • As appropriate, business risks should be escalated up the chain to the appropriate level • Can come from areas such as program dependencies; resources (funding, schedule delivery requirements, people, facilities, suppliers, tools, etc.); priorities; regulations; stakeholders (user community, acquisition officials, etc.); market factors; and weather

SOURCE: DoD, 2017, pp. 22, 79.

Interviewees and Literature Connected Accelerated Acquisition Strategies and Tactics to Potential Increased Program Risks

As part of the interviews, we asked the SMEs to consider how rapid acquisition practices may increase the potential for negative outcomes. We provide some summary examples based on our discussions in Table C.2. For example, interviewees stressed the importance of narrowing requirements sufficiently from the outset to be able to succeed at creating a prototype within the MTA Pathway. If requirements are not sufficiently narrow at the outset, then the schedule goal (which is fixed by law to under five years) may not be met. In other words, it is better to narrow requirements at the start than make performance trade-offs later under schedule pressure. On the other hand, narrow requirements may only satisfy one customer (e.g., SPACECOM) and not address the needs of the joint force. Additionally, narrow requirements may not incorporate longer-term needs and only prioritize shorter-term needs. Interviewees also discussed the importance of using mature technology at higher technology readiness levels during rapid acquisition, to avoid a schedule being negatively affected by difficulties with the technology.

There was a lot of discussion about integration and synchronization. Both are problems that the space acquisition community has grappled with for decades. These are concerns with rapid acquisition. For example, prototypes are being created within the MTA Pathway or Software Acquisition Pathways that must then be integrated into a larger program. There is risk of negative outcomes if these pieces are not properly planned for or addressed up front. The same may be true for cyber considerations.

Likewise, interviewees also discussed programmatic and business events that may lead to increased program risks. For example, given schedule pressures, there is concern that there may not be sufficient time to understand the market, contractor environment, or supply chain risks prior to letting a contract. DoD does not want unexpected surprises (e.g., a contractor in bankruptcy proceedings or other legal proceedings) that may derail its schedule.

Another example is having adequate preparations for sustainment, so that the user community is ready to accept the item without delaying its use during operations. Training must be set up and completed. This same example can be linked to a business event where the user community must submit a POM for the funding for operations and maintenance early enough to plan for fielding.

Interviewees expressed a lot of concerns about resources. Given the uncertainty with obtaining and keeping funding, interviewees mentioned that unexpected changes to funding could throw the program off schedule and that not being able to secure funding at the beginning can lead to negative outcomes.

There was also some concern that organizational culture of risk-aversion in the space community will continue to negatively affect schedule outcomes, and that there may not be sufficient, highly skilled staff to manage the rigors of complicated programs under significant schedule pressures.

Table C.2. Interviewees and Literature Connected Accelerated Acquisition Strategies and Tactics to Potential *Increased* Program Risks

	Acquisition Processes and Functions	Accelerated Acquisition Strategies and Tactics and Potential Negative Impact on Program (i.e., how this may increase risk)
Technical	Requirements	• Requirements are not sufficiently narrow to acquire the product within mandated schedule • No enterprise-wide coordination on prototype requirements
	Research and development	• None identified
	Engineering	• Newer technology may fail during compressed schedule • The USSF does not have a lot of experts in software development • Hiring support contractors for government program teams is also difficult because of the cost of living in some locations • Security clearances are also difficult to get for someone with foreign connections

Acquisition Processes and Functions	Accelerated Acquisition Strategies and Tactics and Potential Negative Impact on Program (i.e., how this may increase risk)
Test and evaluation	• Need to prove that technology will be operational during allowable schedule or make difficult performance trade-offs • There's currently no national test and training range for space systems that have to operate in a contested domain • The test community does not always have the required expertise available to accomplish required testing. Programs have had to request temporary assignment of required experts to accommodate testing requirements and schedules • If testing is done very quickly, safety data may not be shared with the rest of the enterprise or with the engineering community at large
Production, quality, and manufacturing	• None identified
Integration and synchronization	• Space programs involving external launch do not have control of launch schedule • Current architecture and standards may not support modular design and digital engineering needed • Government may need to be integrator for MTA, which is a role that government has not taken on consistently • Alignment/integration across the segments managed by different contractors, which may lead to synchronization problems despite compressed schedule
Cyber/ Intelligence	• Insufficient time or facilities to fully test for cyber considerations in "operational" environment • Program offices may not have an organic testing capability to address cybersecurity • Working in a Special Access Program (SAP) environment precludes conducting acquisition intelligence activities related to threats to the supply chain or cybersecurity • Contested space domain is a new area for the operations, intelligence, and acquisition communities • Insufficient number of embedded intelligence personnel • Depending on other external organizations (e.g., National Security Agency) for support may be a challenge when keeping pace with rapid or urgent acquisition programs

	Acquisition Processes and Functions	Accelerated Acquisition Strategies and Tactics and Potential Negative Impact on Program (i.e., how this may increase risk)
Programmatic	Contract administration	• Working with traditional contractors who have traditionally been responsive to perfect performance incentives rather than to accelerated schedules • Need sufficient time to understand market before engaging industry • Need to secure intellectual property rights early on • Need to develop plan for transition, operations, and maintenance • Need to include cyber resiliency requirements in contract
	Training and education	• Currently no capability for a national test and training range exists to ensure space systems can operate in a contested environment
	Transition to fielding/ sustainment	• User acceptance of less than 100 percent capability critical for success of accelerated program • Early involvement of PMs to transition prototypes to programs of record is not available • Lack of up front planning and documentation in preparation for transition to programs of record • The Systems Program Office controls the SAP clearances. Operational units often do not have SAP clearances and, therefore, access to the acquisition programs they should be supporting
Business	Acquisition process (AAF)	• Incomplete documentation can result in knowledge gaps within the acquisition community • There are different acquisition approaches for hardware (space) and software (ground) intensive systems. The former segment may be acquired through traditional means, while the latter may be acquired using the Software Acquisition Pathway. Additionally, ground and space segments are managed by different contractors, resulting in further misalignment of delivery schedules and products available for integration
	Resources	• Lack of stable budget can inhibit long-term goals such as capability sustainment • Funding is not available on required schedule • Funding is removed from effort during compressed schedule • Less funding is available for smaller efforts • User needs to submit a POM for the operations and maintenance funding • Congress may deny above-threshold programming requests to cover urgent needs • Congress sometimes has difficulty understanding what software programs are buying, which can manifest in future budget cuts • Initial funding cut resulted in longer time frame to award contract • Sequential budget cuts that happen annually cause programs using Rapid Prototyping Authority to lose the time they gained in the first place to start the project faster under such authority
	Industrial base and supply-chain management	• Supply chain security may be difficult to achieve during compressed timelines • Parts may not be acquired during compressed schedule

Acquisition Processes and Functions	Accelerated Acquisition Strategies and Tactics and Potential Negative Impact on Program (i.e., how this may increase risk)
Organizational structure and culture	• MTAs come with more oversight than programs envisioned • Many organizations chartered to support different problems that are also working on different highly classified systems, which decreases communication • Insufficient personnel to staff rapid programs • Organizational culture is still risk averse therefore prioritizing performance over schedule • Involvement of, and poor engagement with, numerous stakeholders can result in technical issues and program delays • Misalignment with supporting functions and organizations • User participation in the acquisition program raises questions about who should be responsible for funding such participation thus lowering the possibility of acquisition programs benefitting from such participation
Leadership support	• Lack of standards and shared vision

SOURCE: RAND discussions with SMEs.

Interviewees and Literature Also Connected Accelerated Acquisition Strategies and Tactics to Potential Decreased Program Risks

We also provide ways that interviewees discussed how accelerated acquisition strategies and tactics have the potential to decrease program risks. We provide some examples in Table C.3.

Defining an 80 percent solution and/or using a narrow set of requirements from the start has resulted in positive cost, schedule, and performance outcomes in prior programs. Likewise, incorporating real-time operational feedback to generate requirements has resulted in better outcomes in programs documented over the past 20 years. Using COTS or heritage technology tends to reduce development time, which helps maintain a positive schedule outcome.

William Roper, the former Assistant Secretary of the Air Force for Acquisition, stressed the importance of applying digital engineering to rapid prototypes as a means of also improving schedule outcomes.[73] Conducting testing in parallel with other parts of the acquisition process and meeting with testing representatives early (including DOT&E) while planning a program have also helped improve schedule outcomes. Finally, urgent acquisition lessons learned include using common interfaces that industry uses to rapidly upgrade the system.

We also identified some additional programmatic and business-related streamlining practices that may improve outcomes. For example, employing a business intelligence function within the contracting office will likely allow the programs to better understand the current state of the industrial base. This function meets regularly with peers across industry. Also, there are multiple proven methods in contracting that will improve outcomes including the use of OTAs and FAR Part 16.5 (IDIQ). Competition with multiple vendors (FFP) during MTA prototyping may also be effective in maintaining cost and schedule.

---

[73] Roper, 2020.

Best practices in urgent acquisition stress the importance of having a small team of highly skilled SMEs available immediately to the acquisition program at the start. This includes embedded functional support and a lean structure with a short and narrow chain of command to eliminate added bureaucratic barriers.

Potentially, having a transition cell in place will help bridge acquisition and operational communities along with liaisons available to interface with the user community. Similarly, having the sustainment infrastructure in place along with the intellectual property rights will help with fielding.

Lining up funding up front, so that program does not rely on reprogramming requests or highly selective prototyping funds, will also be beneficial to program outcomes.

Finally, one of the most important keys to success in rapid acquisition is having leadership support. Leadership must support and create a risk-tolerant environment to improve speed (more acceptability of risk) and also help programs overcome barriers to schedule.

Table C.3. Interviewees and Literature Also Connected Accelerated Acquisition Strategies and Tactics to Potential *Decreased* Program Risks

	Acquisition Processes and Functions	Accelerated Acquisition Strategies and Tactics and Potential Positive Impact on Program (i.e., how this may decrease risk)
Technical	Requirements	• Define 80% solution using narrow set of requirements • Uses alternative requirements process (validated by SPACECOM/Board of Directors or alternative MTA) • Incorporate real-time operational feedback to generate requirements • Consider reduced initial requirements with delivery by later increments • Cut ties with the legacy program to simplify capability requirements for the follow-on programs • Get requirements from users and bring to PIC to see which PEO can satisfy them faster and cheaper
	Research and development	• Use COTS and heritage technology to reduce development time • Conduct development and design concurrently • Partner and collaborate across services, PEOs, Office of the Secretary of Defense, and industry to find solutions
	Engineering	• Apply digital engineering to rapid prototypes • Use rapid prototyping to burn down technical risk • Use higher technology readiness levels • Maintains a list of specifications and standards to encourage programs to test software being reused to encompass all the ways it's being used
	Test and evaluation	• Bring some testing in-house • Conduct testing in parallel with other parts of acquisition process • Use an independent third-party to determine that the code is usable • Meet with testing representatives early (including DOT&E) while planning program • Educate programs conducting rapid acquisition to involve safety representatives at proper steps in the acquisition program to avoid any surprises later on

	Acquisition Processes and Functions	Accelerated Acquisition Strategies and Tactics and Potential Positive Impact on Program (i.e., how this may decrease risk)
	Production, quality, and manufacturing	• Use COTS and heritage technology to reduce production time • Modify proven technology to reduce production/manufacturing time • Use common interfaces that industry uses to rapidly upgrade the system
	Integration and synchronization	• Focus on building end-to-end capability • Interagency effort to promote integration and synchronization using PIC
	Cyber/ intelligence	• Acquisition intelligence function is embedded in the program office, which helps to mitigate the risk of fielding systems vulnerable to threats, including threats to cybersecurity • Incremental acquisition approach with shorter schedules and smaller capability deployments helps to field capability based on current intelligence; new intelligence or threats can be addressed with follow-on capability deployments (Block 2, 3, etc.).
Programmatic	Contract administration	• Employ business intelligence function • Use OTAs, Multi-award OTA (new), FAR Part 16.5 (IDIQ), FAR Part 12 (COTS), Space Enterprise Consortium • Contracting authority resides in the USSF organization • Competition with multiple vendors (FFP) during MTA prototyping
	Training and education	• Focus on highly skilled/agile workforce for acquisition and support functions • Lean forward on training and education for the workforce (particularly software workforce)
	Transition to fielding/ sustainment	• Transition cell in place to bridge acquisition and operational communities • Liaisons available to interface with the user community • Have sustainment infrastructure and intellectual property rights in place
Business	Acquisition process (AAF)	• Tailoring using traditional acquisition processes prior to AAF • MTA (rapid prototyping for hardware-intensive) • Software Acquisition Pathway and continuing to use agile development for software-intensive
	Resources	• Use Head of Space RCO spending authority • Line up funding up front, so do not rely on reprogramming requests or highly selective prototyping fund
	Industrial base and supply-chain management	• Meet regularly with peers across industry • Have an up-to-date knowledge of what contractors, suppliers, and industrial base can do for rapid acquisition
	Organizational structure and culture	• Use a lean structure with short/narrow chain of command and small workforce • Small teams for each program • Embedded functional support
	Leadership support	• Leadership supports risk-tolerant environment to improve speed (more acceptability of risk) • Leadership supports figuring out innovative ways to move forward • Strong senior leadership support has been vital in making sure that other organizations support rapid and urgent programs' cybersecurity needs

SOURCE: RAND discussions with SMEs.

Appendix D. Background Information on Mission Assurance for Class A–D Missions

This appendix provides additional details on the traditional MA processes for Class A–D missions. Table D.1 lists the characteristics associated with the risk profiles for Class A–D missions. Table D.2 summarizes the MA processes for each class. These tables are excerpts from the Aerospace Corporation's *Mission Assurance Guidelines for A–D Mission Risk Classes*. They are included here for the purpose of making this information readily accessible for the reader.

Table D.1. Risk Profile Characteristics for Class A–D

Characteristic	Class A	Class B	Class C	Class D
Risk Acceptance	Minimum practical	Low risk	Moderate risk	Higher risk
National Significance	Extremely critical	Critical	Less critical	Not critical
Payload Type	Operational	Operational or Demo Op	Exploratory or Experimental	Experimental
Acquisition Costs	Highest life cycle Cost (LCC)	High LCC	Medium LCC	Lowest LCC
Complexity	Very high–high	High–medium	Medium–low	Low–medium
Mission Life	>7 years	≤7 years	≤4 years	<1 year
Cost	High	High–medium	Medium–low	Low
Launch Constraints	Critical	Medium	Few	Few–none
Alternatives	None	Few	Some	Significant
Mission Success	All practical Measures	Stringent/minor Compromises	Reduced mission Assurance standards	Few mission Assurance standards
Typical Contract Type	Cost Plus Award Fee (CPAF)*	CPAF–Firm Fixed Price (FFP)	Cost Plus (CP)–FFP	FFP

SOURCE: Reprinted verbatim from Johnson-Roth, 2011, Table 3, p. 5.
* CPAF for Class A is for first of fleet, not once a production program is in place.

77

Table D.2. Summary of Mission Assurance Processes Associated with Class A–D Missions

Mission Assurance Process	Class A	Class B	Class C	Class D
Design Assurance	• **Contractor:** Full design assurance practices, Test driven verification • **Independent Assessment:** Test-Like-You-Fly (TLYF) exceptions, Manufacturing Flow, Millions of Instructions per Second (MIPS) • **Government:** Full review and approval of all processes and products	• **Contractor:** Full design assurance practices • **Independent Assessment:** TLYF exceptions, Manufacturing Flow, MIPS • **Government:** Review and concurrence on process and products, Audit • **Delta:** Reduction in deliveries and formal approval	• **Delta:** Best Practices based, Funding type programmatic control • **Contractor:** Design assurance practices • **Independent Assessment:** Internal TLYF, MIPs • **Government:** Review and concurrence, Audit	• **Delta:** Developer discretion programmatic control • **Contractor:** Essential design assurance practices to mission • **Government:** Periodic review and approval
Requirements Analysis and Validation	• **Contractor:** Validation of Concept of Operations (CONOPS), user scenarios, system readiness, compliance; Subcontractor approval • **Independent Assessment:** for quality, traceability, mission effectiveness, cost/schedule, mission analysis, verification and validation (V&V) of models and simulations • **Government:** Approval (unit level)	• **Delta:** Reduction in deliveries and formal approval • **Contractor:** Class A plus Assume more of oversight responsibility • **Independent Assessment:** Class A Elements • **Government:** Approval (Unit)	• **Delta:** Best practices based, Funding type programmatic oversight • **Contractor:** Mission validation, V&V • **Independent Assessment:** traceability, effectiveness • **Government:** Approval (System)	• **Delta:** Developer discretion programmatic oversight • **Contractor:** Critical requirements flow down • **Government:** Approval (System)

Mission Assurance Process	Class A	Class B	Class C	Class D
Parts, Materials and Processes	• **Part Quality:** Level 1 • **PMPCB:** Customer voting membership • **Radiation:** RDM 2X lot specific, 4X non lot data, SEE <75Mev/ng/sqcm, slant ray analysis • **Radiation Testing:** <margin • **Material:** Heritage envelope or test qualification • **Material approval:** Formal	• **Part Quality:** Level 2 • **PMPCB:** Customer voting negotiated • **Radiation:** Radiation design margin (RDM) 2X lot specific, 4X non lot data, SEE <75Mev/ng/sqcm • **Radiation Testing:** <margin • **Material:** Heritage envelope or test qualification • **Material approval:** Formal	• **Part Quality:** Level 3 • **PMPCB:** No customer voting • **Radiation:** RDM 2X, SEE <37 Mev/ng/sqcm • **Radiation Testing:** Based on data evaluation • **Material:** Heritage envelope or test/analysis qualification • **Material approval:** Informal	• **Part Quality:** Per parts management plan • **PMPCB:** Less formal • **Radiation:** Scoped to critical design • **Radiation Testing:** Scoped to critical design • **Material:** Parts, Materials and Processes Control Board (PMPCB) acceptance • **Material approval:** Informal
Environmental Compatibility	• Environmental compatibility analysis of orbit, mission life, launch factors, mission scenarios • Mission requirements decomposed into individual program plans • Requirement compliance satisfied through testing • No waivers on key performance parameters • Greatest design margins (qual levels)	• Environmental compatibility analysis same as Class A • Mission requirements decomposed same as Class A • Physical testing balanced with analysis, modeling and simulation • Waivers allowed on less critical requirements • Reduced design margins (protoqual levels)	• Environmental compatibility Vetted for impact to other systems and payloads • Mission requirements decomposed based on contractor best practices • Physical testing only used to satisfy mission critical requirements • Waivers acceptable with justified risk impact to mission success • Reduced design margins (protoqual levels)	• Environmental compatibility driven by primary payloads • Mission requirements decomposed based on prior experience • Testing driven for major requirements or driven by primary payload • Waivers acceptable as per Class C for defined requirements • Minimal design margins

Mission Assurance Process	Class A	Class B	Class C	Class D
Reliability Engineering	**Monitoring/Control:** Comprehensive policy, procedures, monitoring and control processes **System Reliability:** System models hardware and software, performance trending, mission reliability **Design Analysis:** Failure Modes and Effects Analysis (FMEA) flight/ground, mechanism Fault Tree Analysis (FTAs), and full worst case analysis (WCA) **Testing/Screening:** Subassembly/part level qualification and assembly level environmental stress screening (ESS) on volume units **Anomaly Management:** First power application reporting, formal closed loop system	**Monitoring/Control:** Policy, procedures, monitoring and control processes with reduced margin requirements **System Reliability:** Minimum SPFs allowed, key parameter trending **Design Analysis:** Failure Modes and Effects Analysis (FMEA) redundancy boundary, mechanism Fault Tree Analysis (FTAs), and reduce worst case analysis (WCA) for susceptible circuits **Testing:** Subassembly/part level qualification and assembly level environmental stress screening (ESS) on volume units **Anomaly Management:** Negotiated first power application reporting, formal closed loop system	**Monitoring/Control:** Monitoring for product spec compliance **System Reliability:** Single string/selective redundancy, parts count analysis, trending limited **Design Analysis:** Functional Failure Modes and Effects Analysis (FMEA) redundancy boundary, critical mechanism Fault Tree Analysis (FTAs), and reduce worst case analysis (WCA) for high risk designs **Testing:** Reduced margins, critical mission reliability driven **Anomaly Management:** Acceptance reporting, formal closed loop system	**Monitoring/Control:** Monitoring required for personnel safety **System Reliability:** Single string baseline, analysis limited **Design Analysis:** S/C payload Failure Modes and Effects Analysis (FMEA) redundancy boundary, safety critical mechanism Fault Tree Analysis (FTAs), and recommended worst case analysis (WCA) not required **Testing:** Qualification to safety critical items only **Anomaly Management:** Internal capture in nonconformance system
System Safety	**Safety Analysis:** Preliminary hazards assessment (PHA), subsystem hazard analysis (SSHA), system hazard analysis (SHA), software system analysis (SSA), operating and support hazard analysis (OSHA), on-orbit hazard analysis, debris **Safety Risk Assessment:** Hazard likelihood/severity **Mishap Reporting:** Formal mishap investigation and reporting	**Safety Analysis:** PHA, SSHA, OSHA **Safety Risk Assessment:** Same as Class A **Mishap Reporting:** Same as Class A	**Safety Analysis:** PHA, OSHA **Safety Risk Assessment:** Same as Class A **Mishap Reporting:** Same as Class A	**Safety Analysis:** PHA, OSHA **Safety Risk Assessment:** Same as Class A **Mishap Reporting:** Same as Class A

Mission Assurance Process	Class A	Class B	Class C	Class D
Configuration/Change Management	• Formal configuration management • (CM) plans, processes and boards integrated throughout the supplier chain with government approval for baseline/change control and configuration audits	• Same as Class A. Government review at sub/supplier levels may be limited	• CM plan not a deliverable; rely on contractor best practices • Formal configuration management is usually initiated once subsystems are integrated • Software CM is initiated earlier	• Not required; applied at the discretion of the developer using best practices
Integration, Test and Evaluation	• **Integration:** Full standard compliance, interface checkout, full copper path evaluation, high fidelity simulator checkout, in-process screening • **Testing – Requirements Compliance and Validation:** Qualification/proto-qualification, full software validation, operability including redundancy checkout, System test including interfaces, launch support test • **TLYF:** All exceptions documented and approved by the customer • **Evaluation:** Maximum customer engagement	• **Integration:** Full standard compliance, interface checkout, full copper path evaluation, Suitable fidelity simulator checkout, In-process Screening • **Testing – Requirements Compliance and Validation:** Proto-qualification with delta cycles, margins, duration, full software validation, operability including redundancy checkout, System test including interfaces, launch support test • **TLYF:** All exceptions documented and approved by the customer • **Evaluation:** Customer review and approval at system/subsystem level	• **Integration:** Standard compliance with tailoring, interface internal checkout, final integration evaluation, GSE validated simulator checkout, reduced in-process screening • **Testing – Requirements Compliance and Validation:** Proto-qualification new hardware/acceptance heritage with delta cycles, margins, duration, software best practices validation, operability, partial system test including interfaces, launch support test • **Evaluation:** Customer review and approval at system level	• **Integration:** Follows best practices, final integration evaluation, GSE certified simulator checkout • **Testing – Requirements Compliance and Validation:** Safety and compatibility testing, software best practices validation, operability. Verification not validation • **Evaluation:** Customer approval of program plan and review at key milestones
Risk Assessment and Management	• Formal joint risk management plan with multiple RMBs • Active management of residual risk • RMB chaired by contractor with customer active participation • Customer approval of programmatic and technical risks mitigation plans	• Joint risk management planning with contractor lead • Residual Risk kept within risk profile • RMB chaired by contractor with customer participation • Customer monitoring of risk mitigation plans	• Contractor risk management planning with customer concurrence • Residual Risk kept within risk profile • RMB internal to contractor • Customer monitoring mission compliance, not margins	• Contractor risk management planning with customer concurrence • Residual risk kept within risk profile • RMB internal to contractor • Customer monitoring mission compliance, not margins

Mission Assurance Process	Class A	Class B	Class C	Class D
Independent Reviews	• Numerous programmatic and technical reviews • SMEs from customer community and contractor • Full standards compliance for entry and exit criteria • All issues tracked to closure	• Small reduction in programmatic and technical reviews • SMEs from customer community and contractor • Standards compliance for negotiated entry and exit criteria • All issues tracked to closure	• Limited programmatic and technical reviews • SMEs from customer community and contractor • General Standards for compliance review conduction • All issues tracked to closure • Review only for moderate to high risk items	• Few key milestone reviews • Internal review based on contractor standards • Best practice standards • All issues tracked to closure • Review only for high risk items
Hardware Quality Assurance	• Full ISO 9001:2000 and AS9100C compliance • Minimum tailoring • Full set of HQA processes to ensure program meets contract and assures mission success.	• Same as Class A program with the exception that there is less customer oversight in areas such as design review and purchasing documents.	• Greatly reduced customer involvement • Relax processes in purchasing, traceability, verification, and environmental controls • Less frequent audits • First article inspection focused on key design features versus 100% verification	• Greater HQA tailoring focused only on key controls and inspection • Audits not typically performed • Nonconformance handling and product preservation potentially done by program resources other than HQA • No first article inspection
Software Assurance	• Full software/firmware SQA process • Independent assessment by customer and contractor SMEs • Detailed artifact capture/closeout • Statistical Reliability Growth • Software Safety Program • SCCB management • Test witnessing	• Same SQA process as Class A • Independent assessment by contractor with customer audit • Core artifact capture/closeout • Statistical Reliability Growth • Significant hazard Software Safety • SCCB management • Test monitoring	• Contractor SQA process • Heritage reuse model • Critical artifact capture/closeout • Process focused Reliability growth • Major hazard Software Safety • SCCB support • Selective test monitoring	• Contractor SQA process recommended • In-line reviews • Major artifacts • Process focused Reliability growth • Personnel/Interface Hazard Software Safety • SCCB support • Test auditing

82

Mission Assurance Process	Class A	Class B	Class C	Class D
Supplier Quality Assurance	• AS9100 certification at contractor, Tier 1 and Tier 2 • Full flow down of customer requirements • Formal verification of supplier certification and process/activity artifacts • Quality Standards customer driven	• AS9100 certification at contractor and major suppliers with intent verification at lower levels • Tailored flow down of customer requirements • Formal verification of supplier certification and process/activity artifacts with tailoring in QMS continuous improvement programs, and documentation process • Quality Standards combined customer/contractor driven	• AS9100 certification at contractor and major suppliers desirable with self-report allowable • Reliance on supplier best practices • Contractual QA based on minimum product standards • Quality Standards best practice driven	• Contractor meets the intent of AS9100 certification at contractor and verification of QA process at supplier for safety-critical elements • Reliance on PI best judgment of acceptable levels of QA • Only key QA practices required
Failure Review Board	• Strive for root cause, seek to eliminate defects in all sibling hardware and verify effective preventive measures • Formal FRB meetings with customer as voting member • FRB control of investigation • Artifacts well documented • Unverified failure commonly results in worst case change out	• Strive for root cause, seek to eliminate defects in all sibling hardware and verify effective preventive measures • Formal FRB meetings with customer but not as voting member • FRB delegation of investigation to cognizant engineer or supplier but closely monitored • Artifacts well documented • Unverified failure thorough evaluation with worst case change out or contingency planning	• Strive for root cause but with a reduced level of control and rigor • FRB meetings based on contractor best practices with results provided to the customer • FRB investigation led by cognizant engineer and suppliers • Less formal presentation of results • Unverified failure processed per contractor policy with eye to cost	• Focus is on actions to return the hardware to service • Failure investigation team may be limited to cognizant engineer and QA (could include supplier) • Less formal results captured in non-conformance system • Unverified failure monitored

83

Mission Assurance Process

Mission Assurance Process	Class A	Class B	Class C	Class D
Corrective/Preventative Action Board	• Likely to have a program specific C/PAB especially for multiple vehicle programs • Same processes as for wide area C/PABs • Programs generate data to support actions to investigate and correct problems • Routine reporting to customer	• Rare to have program unique C/PAB • Programs support wider area C/PABs at company level • Programs generate data used to identify systemic issues or take actions directed by C/PAB • Customer reporting of actions impacting program	• No program unique C/PAB • Programs support wider area C/PABs at company level • Programs generate data used to identify systemic issues or take actions directed by C/PAB • Customer reporting of actions impacting program	• No program unique C/PAB • Programs support wider area C/PABs at company level • Programs generate data used to identify systemic issues or take actions directed by C/PAB • Customer reporting of actions impacting program • Process may be ad hoc for academic and research communities
Alerts, Information Bulletins	• Alerts/Bulletins assessed as potential risks and mitigate to program risk posture • Review of as-design/built, in-line screens, impacts • Supplier same rigor • Regular customer status	• Alerts/Bulletins assessed as potential risks and mitigate to program risk posture • Review similar to Class A but dictated by company policy • Low risk use-as-is • Supplier reporting on impact • Customer status on impact	• Alerts/Bulletins assessed as potential risks and mitigate to program risk posture • Review same as Class B • Moderate risk use-as-is • Supplier responsibility or contractor performs • Only compliance reporting	• Alerts/Bulletins assessed as potential risks and mitigate to program risk posture • Review same as Class B • Moderate risk use-as-is • Contractor performs • Only compliance reporting

SOURCE: Reprinted verbatim from Johnson-Roth, 2011, Table 4, pp. 14–19.

Appendix E. Monthly Acquisition Report Risk Reporting and Analysis

Our principal analysis of risks linked to streamlined acquisition methods relied on a labor-intensive set of interviews with acquisition PMs and other SMEs and could not directly link risks to specific streamlined programs. Thus, we sought to determine whether it would be possible to use routinely collected program data to directly associate programs with specific acquisition challenges and risks. To explore this question, we turned to the Monthly Acquisition Reports (MARs) to review how risks are being reported by PMs and PEOs.

Air Force Instruction 63-101/20-101 requires that PMs prepare MARs for all research, development, test, and evaluation programs with funding greater than $30 million and/or $50 million in procurement over the life of the program. This required reporting includes an assessment of risks. Within the MARs, programs report the top ten issues that they are facing, as well as the possible impact of said issues on their programs, and how they plan to mitigate their impact. This represents a potentially useful source of data for leadership on risks and could be used to compare streamlined and traditional programs to see whether they perceive different risks in their programs.

Approach and Data Collection for the Analysis

For our analysis we collected all DAF reporting programs with MARs, from August 2020 to May 2021. We categorized each of the top reported issues in accordance with the DoD Risk Assessment Framework and the defined technical, programmatic, and business events (as shown in Figure 3.2) to determine the main sources of risk as reported by programs. In addition, the acquisition functions described in Text Box 2.1 can also be binned in accordance with the above described events. Figure E.1 shows how this is accomplished, and Table C.3 has examples.

Figure E.1. Mapping of Acquisition Functions into Technical, Programmatic, and Business Events

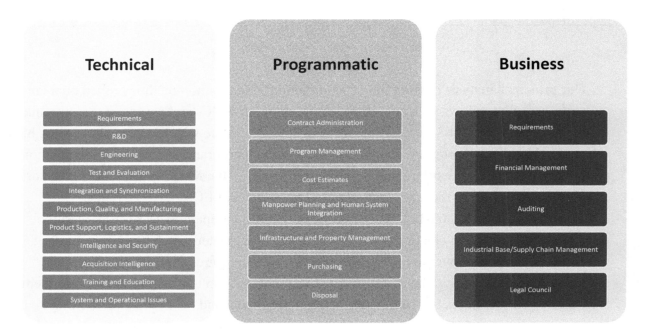

SOURCE: DoD, 2017, pp. 22, 79.

MARs are used for regular review of issues and risks faced at the program level. Information on risks in the MARs is consistent with the DoD Risk, Issue, and Opportunity Management Guide,[74] which underscores the importance of expanding program review beyond risk management, to include issue mitigation management.

The MAR "top issues" section describes the top ten issues identified by the program office, their potential impact on the program, and a description of the mitigation strategy to be adopted. Additionally, each issue is categorized as being a funding, cost, performance, or schedule issue; issues can, and are often categorized in more than one of these categories.

The analysis consisted of MAR data collected across a ten-month period, from August 2020 to May 2021. However, through collecting the data, we noticed that it was often the case that the same issue was reported throughout the entire ten-month period for each of the programs. This leads to the conclusion that programs usually experience a limited number of issues at a time, and these issues usually take many months to resolve.

Data Challenges

This analysis revealed some challenges with using the MAR data to understand risks and to track them over time. A primary challenge was that some MTA programs lacked MAR data, thus

---

[74] DoD, 2017, p. 40.

curtailing our attempt to compare MTA programs with non-MTA programs within SSC. Three of the six SSC MTA programs reported no risks, and three others had their risk information hidden to observers outside SSC for the period of our analysis.

Another of the main challenges associated with the MAR is that risk is described in multiple places throughout the report and needed to be tracked and compiled for our analysis. The reporting template contains a risk section, where the description and type of risk are recorded using a risk matrix and where the PM describes the Likelihood and Consequence of each risk. However, the risk section is rarely completed for non–ACAT I programs. Consistent completion of this risk section could provide a standard method for PMs to discuss risk over time across programs while also creating a structured dataset for easy extraction and analysis for leadership.

Findings

In Figure E.2 we compare the number of unique issues reported for both MTA and non-MTA programs, which are described on the MARs as "804 MTA" and "Acquisition," respectively. The issues are grouped into the DoD three major risk categories business, programmatic, and technical. A total of 27 SSC programs were analyzed, of which three were MTA. The USSF currently has six MTA programs (Next-Gen OPIR, PTS, PTES, ESS, FORGE, MGUE Inc 2 MSI) but only three (PTS, ESS, and Next-Gen OPIR) were available for our viewing. The data showed that although both MTA and non-MTA programs have about the same percentage of business risks, available MTA programs had no technical risk, and programmatic risk comprised a much bigger portion of reported issues in MTA programs than non-MTA ones. As data is only available for three SSC MTA programs this figure does not provide a comprehensive comparison between MTA and not non-MTA programs. No test for statistical significance has been performed for this analysis and we are only making apparent observations.

Figure E.2. Percentage of Unique Issues Reported in MARs for SSC Programs by DoD Risk, Issue, and Opportunity Management Guide Categories

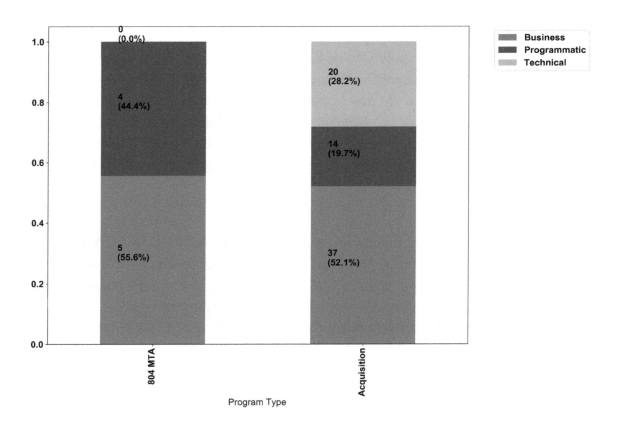

In Figure E.3, we count the number of programs that reported issues and which acquisition function (Figure 2.3) those issues fall under (the figure only shows the top three function for each risk category). Most of the SSC programs that we analyzed showed financial management risks, similar to what was reported in the interviews that we conducted with SMEs. Financial management is a major issue for both MTA and non-MTA programs. In general, according to details described in the MARs, most financial management issues were funding-related, and most were outside the control of the PMs. They were the result of priority changes within the U.S. government or program shortfalls. Because programs report up to ten top issues on the MAR, it is likely that programs experience issues that fall under more than one acquisition function; therefore, if we add all the values in the chart, it will exceed the number of programs.

In addition, the effects of the COVID-19 pandemic were captured by the risks reported by some of the programs: These programs experienced manpower shortages (because of stay-at-home orders, increased production time due to social distancing, or even the lack of material inspectors due to quarantining) and/or industrial base issues related to supply chains that were affected by shutdowns or reduced production.

Figure E.3. Number of SSC Programs That Have Experienced Issues per Acquisition Function

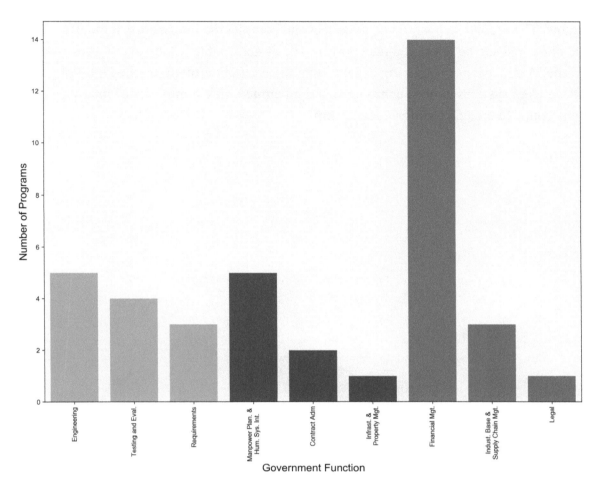

NOTES: Cyan = technical events; brown = programmatic events; blue = business events.

Summary

This exploratory analysis shows that MAR data include the collection of relevant risks that might have an impact on system delivery and thus on MA of space capabilities. MARs are filled out by PMs, who have other ways of informing leadership about risks, so it is not the only vector for transmitting information. However, it does offer a formal approach to collecting risks over time and could be used by leadership as a summary source of information about possible program risks, for tracking risk in individual programs over time, and for portfolio risk analysis.

However, we found some problems with data reporting, including lack of completeness in the descriptions in the MAR risk section. Thus, the data as they currently stand lack sufficient granularity to use as an analytical tool. If the ability to track and understand program risks by portfolio regularly using a standardized source of data would be useful, SSC leadership guidance would be necessary to ensure that the risk sections are completed such that the information can be identified, extracted, and used for analyses.

The first step in addressing these issues would be to provide PMs with a template to identify and prioritize known risks and to continue seeking to identify unknown risks throughout the programs. The second step would be to collect and compile the information regularly. This would allow risks to be appraised on a portfolio basis and enable leadership to track risks trends over time. Also, because delivering capability to the warfighter often requires multiple interdependent systems to be synchronized and integrated, risk summaries of interdependent systems could be created to inform leadership.

Abbreviations

AAF	Adaptive Acquisition Framework
ACAT	Acquisition Category
C2	command and control
COA	course of action
COTS	commercial off-the-shelf
DAF	Department of the Air Force
DevSecOps	development, security, and operations
DoD	U.S. Department of Defense
DOT&E	Director, Operation Test and Evaluation
DT	developmental testing
ESS	Evolved Strategic SATCOM
FAR	Federal Acquisition Regulation
FFP	firm fixed price
FFRDC	federally funded research and development center
FORGE	Future Operationally Resilient Ground Evolution
GAO	U.S. Government Accountability Office
GPS	Global Positioning System
IDIQ	indefinite delivery, indefinite quantity
MA	mission assurance
MAR	Monthly Acquisition Report
MDAP	major defense acquisition program
ME	mission engineering
MGUE	Military GPS User Equipment
MSI	Miniature Serial Interface
MTA	Middle Tier of Acquisition
OPIR	Overhead Persistent Infrared
OT	operational testing
OTA	Other Transaction Authority
PEO	program executive office
PIC	Program Integration Council
PM	program manager
POM	Program Objective Memorandum
PTES	Protected Tactical Enterprise Service
PTS	Protected Tactical SATCOM
RCO	Rapid Capabilities Office
SATCOM	satellite communications
SME	subject-matter expert
SoS	system of systems
SPACECOM	Space Command
Space RCO	Space Rapid Capabilities Office
SSC	Space Systems Command
USSF	U.S. Space Force

References

1st Chief of Space Operations, *Chief of Space Operations' Planning Guidance*, November 9, 2020. As of September 24, 2021:
https://media.defense.gov/2020/Nov/09/2002531998/-1/-1/0/CSO%20PLANNING%20GUIDANCE.PDF

Aerospace Corporation, *Pre-Contract Award Study Schedule Study*, TOR-2016-01191, 2016.

Aerospace Corporation, *Why Does It Take So Long?* TOR-2018-00183, 2018.

Air Force Instruction 63-101/120-101, *Integrated Life Cycle Management*, Department of the Air Force, June 30, 2020.

Air Force Space Command Instruction 10-605, *Operational Acceptance Process*, 2016.

Anton, Philip S., Megan McKernan, Ken Munson, James G. Kallimani, Alexis Levedahl, Irv Blickstein, Jeffrey A. Drezner, and Sydne Newberry, *Assessing Department of Defense Use of Data Analytics and Enabling Data Management to Improve Acquisition Outcomes*, Santa Monica, Calif.: RAND Corporation, RR-3136-OSD, 2019. As of September 8, 2021:
https://www.rand.org/pubs/research_reports/RR3136.html

Anton, Philip S., Brynn Tannehill, Jake McKeon, Benjamin Goirigolzarri, Maynard A. Holliday, Mark A. Lorell, and Obaid Younossi, *Strategies for Acquisition Agility: Approaches for Speeding Delivery of Defense Capabilities*, Santa Monica, Calif.: RAND Corporation, RR-4193-AF, 2020. As of September 8, 2021:
https://www.rand.org/pubs/research_reports/RR4193.html

Brady, Sean, Office of the Deputy Assistant Secretary of Defense for Acquisition Enablers, "Faster Is Possible: DoD Publishes New Software Acquisition Policy," Defense Acquisition University, October 8, 2020. As of September 24, 2021:
https://www.dau.edu/News/Faster-is-possible--DoD-Publishes-New-Software-Acquisition-Policy

Braun, Barbara, Lisa A. Berenberg, Sabrina L. Herrin, Riaz S. Musani, and Douglas A. Harris, *A Class Agnostic Mission Assurance Approach*, Aerospace Corporation, TOR-2021-00133, January 15, 2021.

Broad, William J., and David E. Sanger, "Flexing Muscle, China Destroys Satellite in Test," *New York Times*, January 18, 2007.

Burbey, Douglas, Mindy Gabbert, and Kathryn Bailey, "A Happy Medium: Middle-tier Acquisition Authority Features Flexible Prototype and Fielding Options," *Army ALT Magazine*, September 5, 2019.

Burns, Robert, "US Accuses Russia of Testing Anti-Satellite Weapon in Space," *Washington Post*, July 23, 2020.

Camm, Frank, Thomas C. Whitmore, Guy Weichenberg, Sheng Tao Li, Phillip Carter, Brian Dougherty, Kevin Nalette, Angelena Bohman, and Melissa Shostak, *Data Rights Relevant to Weapon Systems in Air Force Special Operations Command*, Santa Monica, Calif.: RAND Corporation, RR-4298-AF, 2021. As of February 16, 2022:
https://www.rand.org/pubs/research_reports/RR4298.html

Coale, Col. G. Scott, and George Guerra, "Transitioning an ACTD to an Acquisition Program: Lessons Learned from Global Hawk," *Defense AT&L*, September–October 2006. As of December 31, 2020:
https://www.dau.edu/library/defense-atl/DATLFiles/2006_09_10/coa_so_06.pdf

Congressional Budget Office, "CBO Memorandum: The Department of Defense's Advanced Concept Technology Demonstrations," September 1998.

Defense Acquisition University, "Acquisition Category," Acquipedia, undated. As of September 20, 2021:
https://www.dau.edu/acquipedia/pages/ArticleContent.aspx?itemid=313

Defense Acquisition University, "Adaptive Acquisition Framework Pathways," webpage, undated. As of September 11, 2021:
https://aaf.dau.edu/aaf/aaf-pathways/

Defense Acquisition University, "Middle Tier of Acquisition (MTA): MTA Tips," webpage, undated. As of September 24, 2021:
https://aaf.dau.edu/aaf/mta/tips/

Defense Acquisition University, "Software Acquisition," webpage, undated. As of September 22, 2021:
https://aaf.dau.edu/aaf/software/

Defense Acquisition University, Powerful Examples Library, "Powerful Example: USSOCOM Supporting the Hyper Enabled Operator with Agile Logistics Capabilities," webpage, May 6, 2019. As of September 24, 2021:
https://www.dau.edu/powerful-examples/Blog/Powerful-Example-USSOCOM

Defense Acquisition University, Powerful Examples Library, "Powerful Example: B-52 Commercial Engine Replacement Program - Breaking Down Silos to 'Go Faster with Rigor,'" webpage, July 23, 2019. As of September 24, 2021: https://www.dau.edu/News/Powerful-Example--B-52-Commercial-Engine-Replacement-Program

Defense Acquisition University, Powerful Examples Team, "Powerful Example: JRAC helps Warfighters Overcome Urgent Threat From Enemy Drones," August 6, 2019.

Defense Visual Information Distribution Service, "Adaptive Acquisition Framework and Software Pathway," video, January 29, 2020. As of September 24, 2021: https://www.dvidshub.net/video/739994/adaptive-acquisition-framework-and-software-pathway

Department of Defense Directive 5000.01, *The Defense Acquisition System*, September 9, 2020.

Department of Defense Instruction 5000.02, *Operation of the Adaptive Acquisition Framework*, January 23, 2020.

Department of Defense Instruction 5000.80, *Operation of the Middle Tier of Acquisition (MTA)*, December 30, 2019.

Department of Defense Instruction 5000.81, *Urgent Capability Acquisition*, December 31, 2019.

Department of Defense Instruction 5000.87, *Operation of the Software Acquisition Pathway*, October 20, 2020.

DoD—*See* U.S. Department of Defense.

Engert, Pamela A., and Zachary F. Lansdowne, *Risk Matrix User's Guide*, Version 2.2, MITRE Corporation, November 1999. As of September 25, 2021: http://www2.mitre.org/work/sepo/toolkits/risk/ToolsTechniques/files/UserGuide220.pdf

Erwin, Sandra, "Space Force, DoD Agencies, NRO Try to Get on the Same Page on Future Acquisitions," *Space News*, September 22, 2020. As of January 27, 2022: https://spacenews.com/space-force-dod-agencies-nro-try-to-get-on-the-same-page-on-future-acquisitions

Farmer, Jim, "Hidden Value: The Underappreciated Role of Product Support in Rapid Acquisition," *Defense AT&L*, product support issue, March–April 2012.

GAO—*See* U.S. Government Accountability Office.

Holmes-Terry, Shannon, Director, NCR Integration Office, Space Rapid Capabilities Office, "Space Rapid Capabilities Office (SpRCO) Overview," Headquarters U.S. Space Force, October 14, 2020.

Johnson-Roth, Gail, *Mission Assurance Guidelines for A–D Mission Risk Classes*, Aerospace Corporation, TOR-2011(8591)-21, June 3, 2011.

Joint Publication 3-14, *Space Operations*, Joint Chiefs of Staff, April 10, 2018, incorporating Change 1, October 26, 2020.

Judson, Jen, "30 Years: MRAP—Rapid Acquisition Success," *DefenseNews*, October 25, 2016. As of December 31, 2020:
https://www.defensenews.com/30th-annivesary/2016/10/25/30-years-mrap-rapid-acquisition-success/

Kan, Shirley, *China's Anti-Satellite Weapon Test*, Congressional Research Service, RS22652, April 23, 2007. As of September 24, 2021:
https://sgp.fas.org/crs/row/RS22652.pdf

Kim, Yool, Elliot Axelband, Abby Doll, Mel Eisman, Myron Hura, Edward G. Keating, Martin C. Libicki, Bradley Martin, Michael E. McMahon, Jerry M. Sollinger, Erin York, Mark V. Arena, Irv Blickstein, and William Shelton, *Acquisition of Space Systems, Volume 7: Past Problems and Future Challenges*, Santa Monica, Calif.: RAND Corporation, MG-1171/7-OSD, 2015. As of August 15, 2022:
https://www.rand.org/pubs/monographs/MG1171z7.html

Mayer, Lauren A., Mark V. Arena, Frank Camm, Jonathan P. Wong, Gabriel Lesnick, Sarah Soliman, Edward Fernandez, Phillip Carter, Gordon T. Lee, *Prototyping Using Other Transactions: Case Studies for the Acquisition Community*, Santa Monica, Calif.: RAND Corporation, RR-4417-AF, 2020. As of August 16, 2022:
https://www.rand.org/pubs/research_reports/RR4417.html

McKernan, Megan, Jeffrey A. Drezner, and Jerry M. Sollinger, *Tailoring the Acquisition Process in the U.S. Department of Defense*, Santa Monica, Calif.: RAND Corporation, RR-966-OSD, 2015. As of September 8, 2021:
https://www.rand.org/pubs/research_reports/RR966.html

MITRE Corporation, "MITRE Systems Engineering Guide: Risk Management Tools," webpage, undated. As of September 25, 2021:
https://www.mitre.org/publications/systems-engineering-guide/acquisition-systems-engineering/risk-management/risk-management-tools

MITRE Corporation, *Systems Engineering Guide*, 2014.

National Aeronautics and Space Administration Office of Safety and Mission Assurance, *Risk Classification for NASA Payloads*, NPR 8705.4A, April 29, 2021. As of September 2, 2021:
https://nodis3.gsfc.nasa.gov/displayDir.cfm?t=NPR&c=8705&s=4A

Office of the Assistant Secretary of Defense for Homeland Defense and Global Security, *Space Domain Mission Assurance: A Resilience Taxonomy*, September 2015.

Pawlikowski, Ellen, Doug Loverro, and Tom Cristler, "Disruptive Challenges, New Opportunities, and New Strategies," *Strategic Studies Quarterly*, Spring 2012, pp. 27–54.

Project Smart, "The Standish Group Reports: CHAOS," reprinted in 2014 with permission from The Standish Group, 1995.

Public Law 107-314, Bob Stump National Defense Authorization Act for Fiscal Year 2003, December 2, 2002.

Public Law 114-92, National Defense Authorization Act for Fiscal Year 2016, November 25, 2015.

Public Law 116-92, National Defense Authorization Act for Fiscal Year 2020, December 20, 2019.

Romano, Tony, and Jim Whitehead, "Powerful Example: Army IVAS [Integrated Visual Augmentation System] Brings Together the Right Requirements With the Right Acquisition Strategy," Defense Acquisition University, webpage, January 27, 2020. As of September 24, 2021:
https://www.dau.edu/powerful-examples/Blog/Powerful-Example--Army-IVAS

Roper, Will, *There Is No Spoon: The New Digital Acquisition Reality*, October 7, 2020. As of February 14, 2022:
https://software.af.mil/wp-content/uploads/2020/10/There-Is-No-Spoon-Digital-Acquisition-7-Oct-2020-digital-version.pdf

Shelton, William, Cynthia R. Cook, Charlie Barton, Frank Camm, Kelly Elizabeth Eusebi, Diana Gehlhaus, Moon Kim, Yool Kim, Megan McKernan, Sydne Newberry, and Colby P. Steiner, *A Clean Sheet Approach to Space Acquisition in Light of the New Space Force*, Santa Monica, Calif.: RAND Corporation, RR-A541-1, 2021. As of September 24, 2021:
https://www.rand.org/pubs/research_reports/RRA541-1.html

Space and Missile Systems Center Public Affairs, "Operational Acceptance for Space C2 MINERVA," July 12, 2021.

Space Rapid Capabilities Office, homepage, undated. As of September 11, 2021:
https://www.kirtland.af.mil/Units/Space-Rapid-Capabilities-Office

U.S. Department of Defense, *Mission Assurance Strategy*, April 2012.

U.S. Department of Defense, *Department of Defense Risk, Issue, and Opportunity Management Guide for Defense Acquisition Programs*, January 2017.

U.S. Department of Defense, *Department of Defense Fiscal Year (FY) 2021 Budget Estimates: Air Force Research, Development, Test & Evaluation, Space Force*, February 2020.

U.S. Government Accountability Office, *Defense Acquisitions: Rapid Acquisition of MRAP Vehicles*, GAO-10-155T, October 8, 2009.

U.S. Government Accountability Office, *Space Command and Control: Comprehensive Planning and Oversight Could Help DoD Acquire Critical Capabilities and Address Challenges*, GAO-20-146, October 2019.

U.S. Government Accountability Office, *Defense Acquisitions Annual Assessment: Drive to Deliver Capabilities Faster Increases Importance of Program Knowledge and Consistent Data for Oversight*, Report to Congressional Committees, GAO-20-439, June 2020.

U.S. Government Accountability Office, *GPS Modernization: DOD Continuing to Develop New Jam-Resistant Capability, but Widespread Use Remains Years Away*, GAO-21145, January 2021a.

U.S. Government Accountability Office, *DoD Faces Challenges and Opportunities with Acquiring Space Systems in a Changing Environment*, GAO-21-520T, May 2021b.

U.S. Government Accountability Office, *Missile Warning Satellites: Comprehensive Cost and Schedule Information Would Enhance Congressional Oversight*, GAO-21-105249, September 2021c.

U.S. Government Accountability Office, *Space Command and Control: Opportunities Exist to Enhance Annual Reporting*, GAO-22-104685, December 2021d, p. 10.

U.S. Space Force, Space Systems Command, "About Space Systems Command," webpage undated. As of February 19, 2022:
https://www.ssc.spaceforce.mil/About-Us/About-Space-Systems-Command

Van Atta, Richard H., R. Royce Kneece, Jr., and Michael J. Lippitz, *Assessment of Accelerated Acquisition of Defense Programs*, Institute for Defense Analyses, P-8161, 2016.

Williams, Shara, Jeffrey A. Drezner, Megan McKernan, Douglas Shontz, and Jerry M. Sollinger, *Rapid Acquisition of Army Command and Control Systems*, Santa Monica, Calif.: RAND Corporation, RR-274-A, 2014. As of August 16, 2022:
https://www.rand.org/pubs/research_reports/RR274.html

Lightning Source UK Ltd.
Milton Keynes UK
UKHW051930060223
416584UK00013B/283